NANOTECHNOLOGY SCIENCE AND TECHNOLOGY SERIES

PHAGE DISPLAY AS A TOOL FOR SYNTHETIC BIOLOGY

NANOTECHNOLOGY SCIENCE AND TECHNOLOGY SERIES

Safe Nanotechnology
Arthur J. Cornwelle
2009. ISBN: 978-1-60692-662-8

National Nanotechnology Initiative: Assessment and Recommendations
Jerrod W. Kleike (Editor)
2009. ISBN: 978-1-60692-727-4

Nanotechnology Research Collection - 2009/2010. DVD edition
James N. Ling (Editor)
2009. ISBN: 978-1-60741-293-9

Nanotechnology Research Collection - 2009/2010. PDF edition
James N. Ling (Editor)
2009. ISBN: 978-1-60741-292-2

Strategic Plan for NIOSH Nanotechnology Research and Guidance
Martin W. Lang
2009. ISBN: 978-1-60692-678-9

Safe Nanotechnology in the Workplace
Nathan I. Bialor (Editor)
2009. ISBN: 978-1-60692-679-6

Nanotechnology in the USA: Developments, Policies and Issues
Carl H. Jennings (Editor)
2009. ISBN: 978-1-60692-800-4

Nanotechnology: Environmental Health and Safety Aspects
Phillip S. Terrazas (Editor)
2009. ISBN: 978-1-60692-808-0

New Nanotechnology Developments
Armando Barrañón (Editor)
2009. ISBN: 978-1-60741-028-7

Electrospun Nanofibers and Nanotubes Research Advances
A. K. Haghi (Editor)
2009. ISBN: 978-1-60741-220-5

Electrospun Nanofibers and Nanotubes Research Advances
A. K. Haghi (Editor)
2009. ISBN: 978-1-60876-762-5 (Online Book)

Carbon Nanotubes: A New Alternative for Electrochemical Sensors
*Gustavo A. Rivas, María D. Rubianes, María L. Pedano,
Nancy F. Ferreyra, Guillermina Luque and Silvia A. Miscoria*
2009. ISBN: 978-1-60741-314-1

Polymer Nanocomposites: Advances in Filler Surface Modification Techniques
Vikas Mittal (Editor)
2009. ISBN: 978-1-60876-125-8

Nanostructured Materials for Electrochemical Biosensors
Yogeswaran Umasankar, S. Ashok Kuma and Shen-Ming Chen (Editors)
2009. ISBN: 978-1-60741-706-4

Magnetic Properties and Applications of Ferromagnetic Microwires with Amorphous and Nanocrystalline Structure
Arcady Zhukov and Valentina Zhukova
2009. ISBN: 978-1-60741-770-5

Electrospun Nanofibers Research: Recent Developments
A.K. Haghi (Editor)
2009. ISBN: 978-1-60741-834-4

Nanofibers: Fabrication, Performance, and Applications
W. N. Chang (Editor)
2009. ISBN: 978-1-60741-947-1

Nanofibers: Fabrication, Performance, and Applications
W. N. Chang (Editor)
2009. ISBN: 978-1-61668-288-0 (Online Book)

Bio-Inspired Nanomaterials and Nanotechnology
Yong Zhou (Editor)
2009. ISBN: 978-1-60876-105-0

Nanotechnology: Nanofabrication, Patterning and Self Assembly
Charles J. Dixon and Ollin W. Curtines (Editor)
2010. ISBN: 978-1-60692-162-3

Gold Nanoparticles: Properties, Characterization and Fabrication
P. E. Chow (Editor)
2010. ISBN: 978-1-61668-009-1

Gold Nanoparticles: Properties, Characterization and Fabrication
P. E. Chow (Editor)
2010. ISBN: 978-1-61668-391-7 (Online book)

Micro Electro Mechanical Systems (MEMS): Technology, Fabrication Processes and Applications
Britt Ekwall and Mikkel Cronquist (Editors)
2010. ISBN: 978-1-60876-474-7

Nanomaterials: Properties, Preparation and Processes
Vinicius Cabral and Renan Silva (Editors)
2010. ISBN: 978-1-60876-627-7

Nanopowders and Nanocoatings: Production, Properties and Applications
V. F. Cotler (Editor)
2010. ISBN: 978-1-60741-940-2

Barrier Properties of Polymer Clay Nanocomposites
Vikas Mittal (Editor)
2010. ISBN: 978-1-60876-021-3

Nanomaterials Yearbook - 2009. From Nanostructures, Nanomaterials and Nanotechnologies to Nanoindustry
Gennady E. Zaikov and Vladimir I. Kodolov (Editors)
2010. ISBN: 978-1-60876-451-8

Nanoparticles: Properties, Classification, Characterization, and Fabrication
Aiden E. Kestell and Gabriel T. DeLorey (Editors)
2010. ISBN: 978-1-61668-344-3

Nanoporous Materials: Types, Properties and Uses
Samuel B. Jenkins (Editor)
2010. ISBN: 978-1-61668-182-1

Nanoporous Materials: Types, Properties and Uses
Samuel B. Jenkins (Editor)
2010. ISBN: 978-1-61668-650-5 (Online book)

Mechanical and Dynamical Principles of Protein Nanomotors: The Key to Nano-Engineering Applications
A. R. Khataee and H. R. Khataee
2010. ISBN: 978-1-60876-734-2

TiO2 Nanocrystals: Synthesis and Enhanced Functionality
Ji-Guang Li , Xiaodong Li and Xudong Sun
2010. ISBN: 978-1-60876-838-7

Nanomaterial Research Strategy
Earl B. Purcell (Editor)
2010. ISBN: 978-1-60876-845-5

Magnetic Pulsed Compaction of Nanosized Powders
G.Sh Boltachev, K.A Nagayev, S.N. Paranin, A.V. Spirin, N.B. Volkov
2010. ISBN: 978-1-60876-856-1

Nanostructured Conducting Polymers and their Nanocomposites:
Classification, Properties, Fabrication and Applications
Ufana Riaz and S.M. Ashraf
2010. ISBN: 978-1-60876-943-8

Phage Display as a Tool for Synthetic Biology
Santina Carnazza and Salvatore Guglielmino
2010. ISBN: 978-1-60876-987-2

Bioencapsulation in Silica-Based Nanoporous Sol-Gel Glasses
Bouzid Menaa, Farid Menaa, Carla Aiolfi-Guimarães, and Olga Sharts
2010. ISBN: 978-1-60876-989-6

ZnO Nanostructures Deposited by Laser Ablation
M. Martino, D. Valerini, A.P. Caricato A. Cretí, M. Lomascolo and R. Rella
2010. ISBN: 978-1-61668-034-3

Development and Application of Nanofiber Materials
Shou-Cang Shen, Wai-Kiong Ng, Pui-Shan Chow, Reginald B.H. Tan
2010. ISBN: 978-1-61668-416-7

Development and Application of Nanofiber Materials
Shou-Cang Shen, Wai-Kiong Ng, Pui-Shan Chow, Reginald B.H. Tan
2010. ISBN: 978-1-61668-829-5 (Online book)

Polymers as Natural Composites
Albrecht Dresdner and Hans Gärtner (Editors)
2010. ISBN: 978-1-61668-168-5

Polymers as Natural Composites
Albrecht Dresdner and Hans Gärtner (Editors
2010. ISBN: 978-1-61668-886-8 (Online book)

Synthesis and Engineering of Nanostructures by Energetic Ions
Devesh Kumar Avasthi, Jean Claude Pivin (Editors)
2010. ISBN: 978-1-61668-209-5

From Gold Nano-Particles Through Nano-Wire to Gold Nano-Layers
V. Švorčík, Z. Kolská, P. Slepička, V. Hnatowicz
2010. ISBN: 978-1-61668-316-0

From Gold Nano-Particles Through Nano-Wire to Gold Nano-Layers
V. Švorčík, Z. Kolská, P. Slepička, V. Hnatowicz
2010. ISBN: 978-1-61668-722-9 (Online book)

Phase Mixture Models for the Properties of Nanoceramics
Willi Pabst and Eva Gregorova
2010. ISBN: 978-1-61668-673-4

Phase Mixture Models for the Properties of Nanoceramics
Willi Pabst and Eva Gregorova
2010. ISBN: 978-1-61668-898-1 (Online book)

Applications of Electrospun Nanofiber Membranes for Bio-separations
Todd J. Menkhaus, Lifeng Zhang, Hao Fong
2010. ISBN: 978-1-60876-782-3

NANOTECHNOLOGY SCIENCE AND TECHNOLOGY SERIES

PHAGE DISPLAY AS A TOOL FOR SYNTHETIC BIOLOGY

SANTINA CARNAZZA
AND
SALVATORE GUGLIELMINO

Nova Science Publishers, Inc.
New York

Copyright © 2010 by Nova Science Publishers, Inc.

All rights reserved. No part of this book may be reproduced, stored in a retrieval system or transmitted in any form or by any means: electronic, electrostatic, magnetic, tape, mechanical photocopying, recording or otherwise without the written permission of the Publisher.

For permission to use material from this book please contact us:
Telephone 631-231-7269; Fax 631-231-8175
Web Site: http://www.novapublishers.com

NOTICE TO THE READER

The Publisher has taken reasonable care in the preparation of this book, but makes no expressed or implied warranty of any kind and assumes no responsibility for any errors or omissions. No liability is assumed for incidental or consequential damages in connection with or arising out of information contained in this book. The Publisher shall not be liable for any special, consequential, or exemplary damages resulting, in whole or in part, from the readers' use of, or reliance upon, this material. Any parts of this book based on government reports are so indicated and copyright is claimed for those parts to the extent applicable to compilations of such works.

Independent verification should be sought for any data, advice or recommendations contained in this book. In addition, no responsibility is assumed by the publisher for any injury and/or damage to persons or property arising from any methods, products, instructions, ideas or otherwise contained in this publication.

This publication is designed to provide accurate and authoritative information with regard to the subject matter covered herein. It is sold with the clear understanding that the Publisher is not engaged in rendering legal or any other professional services. If legal or any other expert assistance is required, the services of a competent person should be sought. FROM A DECLARATION OF PARTICIPANTS JOINTLY ADOPTED BY A COMMITTEE OF THE AMERICAN BAR ASSOCIATION AND A COMMITTEE OF PUBLISHERS.

LIBRARY OF CONGRESS CATALOGING-IN-PUBLICATION DATA

Phage display as a tool for synthetic biology / Santina Carnazza and
Salvatore Guglielmino.
     p. ; cm.
  Includes bibliographical references and index.
  ISBN 978-1-60876-987-2 (softcover)
  1. Bacteriophages. 2. Microbial biotechnology. 3. Protein engineering.
  4. Nanotechnology. I. Guglielmino, Salvatore, 1950- II. Title.
  [DNLM: 1. Bacteriophages. 2. Biomimetic Materials--chemical synthesis.
  3. Nanostructures. 4. Nanotechnology. 5. Protein Engineering. QW 161
C288p 2010]
  TP248.27.M53C37 2010
  660.6--dc22
                              2009052739

Published by Nova Science Publishers, Inc. † New York

CONTENTS

Abstract		i
Chapter 1	Synthetic Biology Applications in Nanobiotechnology	1
Chapter 2	The "Never Born Proteins"	5
Chapter 3	Protein Engineering and Directed Evolution	9
Chapter 4	In Vitro and Biological Display Technologies	13
Chapter 5	Phage Display for Directed Molecular Evolution	19
Chapter 6	Antibody Phage	23
Chapter 7	Phage as Bioselective Probes	27
Chapter 8	Phage-Derived Nanomaterials	33
Chapter 9	Phage Perspectives in Nanobiotechnology	41
References		47
Index		71

Chapter 1

SYNTHETIC BIOLOGY APPLICATIONS IN NANOBIOTECHNOLOGY

Synthetic biology studies how to build artificial biological systems for engineering applications, using advanced tools of system design, modeling and simulation, as well as the most recent experimental techniques.

There can be various approaches to synthetic biology: engineering of biological systems; redesigning life, by constructing biological systems, aimed to bridge gaps in our current understanding of biology; creating alternative life, by using unnatural molecules in living systems.

The focus is often on ways of taking parts of natural biological systems, characterizing and simplifying them, and using them as components of novel, engineered, highly unnatural life forms.

The experimental work has a philosophical counterpart, arising in a special way when chemistry, physics and engineering move towards biology.

Biologists are interested in synthetic biology because it provides a complementary perspective from which to consider, analyze, and ultimately understand the living world. Being able to design and build a system is also one very practical measure of understanding. Physicists, chemists and others are interested in synthetic biology as an approach with which to probe the behavior of molecules and their activity inside living cells. Engineers view biology as a technology; they are interested in synthetic biology because the living world provides an apparently rich yet largely unexplored way for controlling and processing information, materials, and energy.

Protein evolution in vitro technologies, together with protein and genetic engineering, can provide the tools needed for rapid design, fabrication, and testing of systems. Studies of cellular function, discovery of new therapeutic targets, and detailed mechanistic and structural analyses of proteins rely on specific binding reagents. Display techniques are powerful tools to generate, select, and evolve such binding reagents completely in vitro, and they have great potential for biotechnological, medical and proteomic applications.

In particular, phage display technology has become a fundamental tool in functional genomics and proteomics, as well as an invaluable component of biotechnology. Specific ligands can be isolated from highly diverse peptide libraries against virtually any target of interest, and successfully used in various research fields. It is used for studying protein-protein, protein-peptide, and protein-DNA interactions; moreover, it aids to explore protein structure/function and it extends to the synthesis of artificial proteins with random sequences.

During the past ten years, phage display has evolved into a well-accepted technology, and in a short time has delivered both sophisticated, high-quality antibody phage and recombinant phage probes for detection of pathogenic agents. Improved library construction approaches —in combination with innovated vector design, display formats and screening methods— have further extended the technology. It seems probable that extremely diverse phage-display libraries contain multiple solutions to most binding problems. Phage display technology aids to explore the links between protein structure and function, and this information will in turn expedite the process of directed molecular evolution; moreover, it extends to the synthesis of artificial proteins with random sequences.

Phage antibodies are likely to play a great role in the generation of analytical reagents and therapeutic drugs, offering major advantages in terms of speed and throughput for research and target identification/validation. The greatest challenge for the future will be to translate our ability to create binding sites with tailored size, affinity, valency and sequence, into antibody molecules with improved clinical benefit.

Recombinant phage selected by phage-display find application also as biosorbent and diagnostic probe in micro- and nano-devices, as effective surrogates of antibodies, used to date. For their properties, phage probes may meet the strong criteria —stability, fastness, sensitiveness, accuracy, and inexpensiveness— for development of bioaffinity sensors for biological monitoring. For example, they may be used for separation and purification of bacteria prior to their identification with polymerase chain reaction, immunoassays, flow cytometry, or other methods. Otherwise, they can be used to develop rapid real-time diagnostic arrays, by themselves recognizing and binding selectively and specifically the target, with no need of further characterization. Highly sensitive and accurate field-usable devices could have a number of applications in biomedical field as well as in environment and food monitoring, and detection of biological warfare agents.

Moreover, important developments have been made in the synthesis of bio-nanostructures with nano-crystals, including protein-shelled viruses modified by metallic or semiconducting nanoparticles. Nanostructured biomaterials represent an ideal system for use in biological detection due to their unique selectivity. Organizing ordered inorganic nanoparticles by using soft materials as templates is essential for constructing electronic devices with new functionalities. The concept of conjugating nanoparticles with biomolecules opens up new possibilities for making functional next-generation electronic devices using biomaterial systems. Phage-display libraries have been successfully used to identify, develop and amplify binding between organic peptide sequences and inorganic metal and semiconductor substrates, in order to provide templates for the synthesis of polynanocrystalline nanowires, nanotubes and nanorings for applications in advanced nanoelectronic devices.

Several other potential applications in the modern biotechnology industry have been recently recognized for recombinant phage –phage-therapy, gene delivery, phage-display vaccination, targeted therapy.

Tailored selection processes, in combination with improved library construction and innovated vector design and display formats, open the door wide for new sophisticated applications of phage display technology in synthetic biology.

Chapter 2

THE "NEVER BORN PROTEINS"

One of the aims of protein engineering is to design proteins from scratch — for example, new protein-based materials and artificial enzymes. The chemical approach to synthetic biology is concerned with the synthesis of molecular structures and/or multi-molecular organized systems –proteins, nucleic acids, vesicular forms, and other– that do not exist in nature.

Some peptide materials have been successfully designed [McGrath et al. 1992; Urry et al. 1995; Deming 1997], as well as de novo peptides with specified folds [Bryson et al. 1995]. One interesting project belonging to this chemical frontier of synthetic biology concerns the so-called "never born proteins" (NBPs), meaning proteins that have not been produced and/or selected by nature in the course of biological evolution [Luisi 2007]. The proteins existing in nature make only an infinitesimal fraction of the theoretically possible structures, and our life is based on a very limited number of structures. This elicits the question why and how the protein structures existing in our world have been selected out, with the underlying question whether they have something very particular from the structural or thermodynamic point of view that made the selection possible. For example, the few structures selected might be the only ones to be stable (i.e., with the correct folding); or water soluble; or those which have very particular viscosity and/or rheological properties. A second point of view is that "our" proteins have no extraordinary physical properties at all; they have been selected by "chance" among an enormous number of possibilities of quite similar compounds, and it happened that they were capable of fostering cellular life. This last belongs to the so-called "contingency" theory.

The NBPs can be produced in laboratory either by chemical synthesis –e.g. fragment condensation of short peptides with selection governed by the contingency of the environmental conditions– or the modern molecular biology techniques –such as the phage display method.

The principle to produce NBPs is simple: if one makes a long string of DNA purely randomly, the probability of hitting an existing sequence in nature is practically zero. If then this DNA is processed by standard recombinant DNA and in vivo expression techniques, a non-existing polypeptide will be obtained, which, when globularly folded, is already a NBP.

In practice, the aim is to produce a very large library of totally random, de novo proteins having no homology with known proteins, and to investigate whether these synthetic biology products are really so different with respect to natural proteins, in terms of stability, solubility, or folding.

Luisi's group [Chiarabelli et al. 2006a, 2006b; De Lucrezia et al. 2006a, 2006b] tackled the question by investigating folding ability of NBPs, considering it a particularly important and stringent criterion, as the prerequisite for the biological activity of proteins determined by their primary structure. The strategy adopted was based on the well-accepted observation that folded proteins are not easily digestible by proteases. It involved the insertion of the tripeptide PRG (proline-arginine-glycine), substrate for the proteolytic enzyme thrombin, in the otherwise totally random protein sequence. In this way, each of the new proteins had the potentiality of being digested by the enzyme, with the expectation, however, that globularly folded NBPs would be protected from digestion. The larger part of the population was rapidly hydrolyzed, but ~ 20% of the NBPs were highly resistant to the action of thrombin, suggesting that folding is indeed a general property, something that arises naturally, even for proteins of medium length. A significant percentage of periodic structure, α-helix in particular, was present, and, furthermore, the globular folding was thermoreversible, indicating to be under thermodynamic control.

It appears possible at this point to state that folding and thermodynamic stability are not properties that are restricted to extant proteins, and that, on the contrary, they appear to be rather common features of randomly created polypeptides. On the basis of this, one is tempted to propose that "our" natural proteins do not belong to a class of polypeptides with privileged physical properties. And, by inference, one could say that this kind of data permit to brake a lance in favour of the scenario of contingency.

Of course, the NBPs may have also bio-technological importance and be very interesting from the structural point of view: they could, for example, display novel catalytic and structural features that have not been observed in natural proteins.

Indeed, a more difficult challenge is achieving novel catalytic function in artificial proteins with an efficacy and specificity similar to that of natural enzymes. Most efforts so far have tended to use the natural combinatorial mechanism of the immune system to develop antibodies with catalytic functions [Tramontano et al. 1986]. Recently, however, advances in computational methods have been exploited to transform a non-catalytic protein receptor into a mimic of a natural enzyme by rationally mutating several residues in the binding site [Dwyer et al. 2004]. Thus, rational protein design does not necessarily have to be conducted wholly de novo: existing protein folds can be used for the "scaffolding", and one can focus simply on retooling the active site. Moreover, proteins can be rationally modified to bind to new, non-natural substrates.

These developments in protein design have been adapted for nanotechnological uses. For example, the versatility of the immune system has been used to generate antibodies that will recognize and bind to fullerenes [Chen et al. 1998], carbon nanotubes [Erlanger et al. 2001], and a variety of crystal surfaces [Izhaky and Addadi 1998]. Synthetic biology, however, could provide the tools and understanding needed to develop "nanobiotechnology" in a more systematic manner, as well as to expand the scope of what it might achieve [Ball 2005].

Synthetic biology includes the broad redefinition and expansion of biotechnology, with the ultimate goals of being able to design and build engineered biological systems that process information, manipulate chemicals, fabricate materials and structures, produce energy, provide food, and maintain and enhance human health and our environment [Chopra and Kamma 2006].

Chapter 3

PROTEIN ENGINEERING AND DIRECTED EVOLUTION

Protein engineering is a relatively young discipline, aimed to develop novel useful or valuable proteins with new and uniquely functional attributes [Graff et al. 2004]. Much research is currently taking place into the understanding of the fundamental rules linking a protein's structure to its function, and it involves the application of science, mathematics and economics.

There are two main strategies for protein engineering. The first is known as rational design, in which detailed knowledge of the structure and function of the protein is used to make desired changes. This has the advantage of being generally inexpensive and easy, since site-directed mutagenesis techniques are well-developed. However, there is a major drawback in that detailed structural knowledge of a protein is often unavailable, and even when it is available, it can be extremely difficult to predict the effects of various mutations.

Computational protein design algorithms seek to identify amino acid sequences that have low energies for target structures. While the sequence-conformation space that needs to be searched is large, the most challenging requirement for computational protein design is a fast, yet accurate, energy function that can distinguish optimal sequences from similar suboptimal ones. Using computational methods, a protein with a novel fold has been designed [Yuan et al. 2005], as well as sensors for unnatural molecules [Arnold 1998].

The second strategy is known as directed evolution. This method mimics natural evolution to evolve proteins with desirable properties not found in nature, and generally produces superior results to rational design. Random mutagenesis is applied to a protein, and a selection regime is used to pick out variants that have the desired qualities. Further rounds of mutation and selection are then applied, in order to allow an increase in functional density of the protein of interest, identifying interesting mutants. It is thus possible to use this method to optimize properties that were not selected for in the original organism, including catalytic specificity, thermostability and many others.

A typical directed evolution experiment involves two steps:

1. Library creation: The gene encoding the protein of interest is mutated and/or recombined at random to create a large library of gene variants.
2. Library screening: The library is screened by the researcher using a high-throughput screen to identify mutants or variants that possess the desired properties. "Winner" mutants identified in this way then have their DNA sequenced to understand what mutations have occurred.

The evolved protein is then characterized using biochemical methods.

The great advantage of directed evolution techniques is that they require no prior structural knowledge of a protein, nor it is necessary to be able to predict what effect a given mutation will have. Indeed, the results of directed evolution experiments are often surprising in that desired changes are often caused by mutations that no one would have expected. The drawback is that they require high-throughput, which is not feasible for all proteins. Large amounts of recombinant DNA must be mutated and the products screened for desired qualities. The sheer number of variants often requires expensive robotic equipment to automate the process. Furthermore, not all desired activities can be easily screened for. New advancements in high-throughput technology will greatly expand the capabilities of directed evolution.

An additional technique known as *DNA shuffling*, or *sexual Polymerase Chain Reaction* (PCR) [Stemmer 1994], mixes and matches pieces of successful variants in order to rapidly propagate beneficial mutations, thus producing better results in a directed evolution experiment. This process mimics recombination that occurs naturally during sexual reproduction and is used to rapidly increase DNA library size.

DNA shuffling is a PCR without synthetic primers. In this process, a family of related genes are first cut with enzymes. The gene fragments then are heated up to separate them into single-stranded templates. Some of these fragments will bind to other fragments that share complementary DNA regions, which in some cases will be from other family members. Regions of DNA that are non-complementary hang over the ends of the templates, and the PCR reaction then treats the complementary regions as primers and builds the new double-helical DNA. But PCR also adds bases to the overhanging piece of the primer, forming a double helix there, too. This ultimately creates a mixed structure or "chimera". In the final step, PCR reassembles these chimeras into full-length, shuffled genes.

Application of these methods to engineer protein cores, active sites and macromolecular interfaces will contribute greatly to our ability to both understand and rationally manipulate the physicochemical properties that drive protein function.

Chapter 4

IN VITRO AND BIOLOGICAL DISPLAY TECHNOLOGIES

One of the most powerful strategies to improve the properties of proteins or even create new ones is to imitate the strategy of evolution in the test tube, through an in vitro iteration between diversification and selection, by means of display technologies. The directed evolution of proteins using display methods can be engineered for specific properties and selectivity. A variety of display approaches are employed for the engineering of optimized human antibodies, as well as protein ligands, for such diverse applications as protein arrays, separations, and drug development.

In vitro display technologies, namely ribosome and mRNA display [for a review, Amstutz et al. 2001; Lipovsek and Plückthun 2004], combine two important advantages for identifying and optimizing ligands by evolutionary strategies. First, by obviating the need to transform cells in order to generate and select libraries, they allow much higher library diversity. Second, by including PCR as an integral step in the procedure, they make PCR-based mutagenesis strategies convenient. The resulting iteration between diversification and selection allows true Darwinian protein evolution to occur in vitro. Successful examples of high-affinity, specific target-binding molecules selected by in-vitro display methods include peptides, antibodies, enzymes, and engineered scaffolds, such as fibronectin type III domains [Koide et al. 1998; Xu et al. 2002] and synthetic ankyrins, which can mimic antibody function [Binz et al. 2003, 2004].

RIBOSOME DISPLAY

Ribosome display is a technique used to perform in vitro protein evolution to create proteins that can bind to a desired ligand. It was first developed by Mattheakis *et al* [1994] for the selection of peptides and further improved for folded proteins [Hanes and Plückthun 1997; He and Taussig 1997, 2007]. A fusion protein is constructed in which the domain of interest is fused to a C-terminal tether, such that this domain can fold while the tether is still in the ribosomal tunnel. This fusion construct lacks a stop codon at the mRNA level, thus preventing release of the mRNA and the polypeptide from the ribosome. The process results in translated proteins that remain associated with ribosome and their mRNA progenitor, which is used, as a non-covalent ternary complex, to bind to an immobilized ligand in a selection step. The mRNA-protein hybrids that bind well are then reverse transcribed to cDNA and their sequence amplified via PCR. The end result is a nucleotide sequence that can be used to create tightly binding proteins.

The complex of mRNA, ribosome, and protein is stabilized with the lowering of temperature and the addition of cations such as Mg^{2+}. During the subsequent panning stages, the complex is introduced to surface-bound ligand in several ways: using an affinity chromatography column with a resin bed containing ligand, a 96-well plate with immobilized surface-bound ligand, or magnetic beads that have been coated with ligand. The complexes that bind well are immobilized. Subsequent elution of the binders, via high salt concentrations, chelating agents, or mobile ligands which complex with the binding motif of the protein, allows dissociation of the mRNA. The mRNA can then be reverse transcribed back into cDNA, and thus, the genetic information of the binding polypeptides is available for analysis, then it can undergo mutagenesis, and iteratively fed into the process with greater selective pressure to isolate even better binders.

By having the protein progenitor attached to the complex, the process of ribosome display skips the microarray/peptide bead/multiple-well sequence separation that is common in assays involving nucleotide hybridization and provides a ready way to amplify the proteins that do bind without decrypting the sequence until necessary. At the same time, this method relies on generating large, concentrated pools of sequence diversity without gaps and keeping these sequences from degrading, hybridizing, and reacting with each other in ways that would create sequence-space gaps.

In addition, as ribosome display is the first method for screening and selection of functional proteins performed completely in vitro, it circumvents many drawbacks of in vivo selection technologies. First, the diversity of the library is not limited by the transformation efficiency of bacterial cells, but only by the number of ribosomes and different mRNA molecules present in the test tube. Second, random mutations can be introduced easily after each selection round, as no library must be transformed after any diversification step. This allows simple directed evolution of binding proteins over several generations.

In ribosome display, the physical link between genotype and phenotype is accomplished by an mRNA–ribosome–protein complex that is used for selection. As this complex is stable for several days under appropriate conditions, several selections can be performed. Ribosome display allows protein evolution through a built-in diversification of the initial library during selection cycles. Thus, the initial library size no longer limits the sequence space sampled.

This technology of directed evolution over many generations is currently being exploited to address fundamental questions of protein structure and stability [Jermutus et al. 2001; Hanes et al. 2000a], catalysis [Amstutz et al. 2002; Cesaro-Tadic et al. 2003], as well as interesting biomedical applications. Recently, the potential of ribosome display for directed molecular evolution was recognized and developed into a rapid and simple affinity selection strategy to obtain scFv fragments of antibodies with affinities in the low picomolar range [Schaffitzel et al. 1999; Hanes et al. 2000b]. The authors selected a range of different scFvs with picomolar affinity from a fully synthetic naïve antibody scFv library using ribosome display. All of the selected antibodies accumulated beneficial mutations throughout the selection cycles. This display method can apply also to other members of the immunoglobulin superfamily; for example single V-domains which have an important application in providing specific targeting to either novel or refractory cancer markers [Irving et al. 2001]. These works demonstrated that ribosome display not only allows the selection of library members but also further evolves them, thereby mimicking the strategy of the immune system itself.

It was also demonstrated that even those proteins can be selected that cannot be expressed at all in vivo [Schimmele and Plückthun 2005; Schimmele et al. 2005].

Ribosome display systems that are well proven, by the evolution of high affinity antibodies and the optimisation of defined protein characteristics, generally use an *Escherichia coli* cell extract for in vitro translation and display of an mRNA library. More recently, a cell-free translation system has been produced by combining recombinant *E. coli* protein factors with purified 70S ribosomes [Villemagne et al. 2006]. Higher cDNA yields are recovered from ribosome display selections when using a reconstituted translation system and the degree of improvement seen is selection specific. These effects are likely to reflect higher mRNA and protein stability and potentially other advantages that may include protein specific improvements in expression. Reconstituted translation systems therefore enable a highly efficient, robust and accessible prokaryotic ribosome display technology.

Competing methods for protein evolution in vitro are mRNA display, yeast display, bacterial display and phage display.

mRNA DISPLAY

Like other biological display technologies, mRNA display technology provides easily accessible coding information for each peptide/protein displayed [Roberts and Szostak 1997; Nemoto et al. 1997]. In mRNA display, mRNA is first translated and then covalently bonded to the nascent polypeptide it encodes, using puromycin as an adaptor molecule. The covalent mRNA–protein adduct is purified from the ribosome and used for selection. Puromycin is an analogue of the 3' end of a tyrosyl-tRNA with a part of its structure mimics a molecule of adenosine, and the other part mimics a molecule of tyrosine. Compared to the cleavable ester bond in a tyrosyl-tRNA, puromycin has a non-hydrolysable amide bond. As a result, puromycin interferes with translation, and causes premature release of translation products. The protein and the mRNA are thus coupled and are subsequently isolated from the ribosome and purified. In the current protocol, a cDNA strand is then synthesized to form a less sticky RNA–DNA hybrid and these complexes are finally used for selection.

mRNA display technology has many advantages over the other display methods. The biological display libraries (phage, yeast and bacterial) have polypeptides or proteins expressed on the respective microorganism's cell surface, and the accompanying coding information for each polypeptide or protein is retrievable from the microorganism's genome. However, the library size for the in vivo display systems is limited by the transformation efficiency of each organism. For example, the library size for phage and bacterial display is limited to 1-10 $\times$ 10^9 different members. The library size for yeast display is even smaller. Moreover, these cell-based display systems only allow the screening and enrichment of peptides/proteins containing natural amino acids. In contrast, mRNA display and ribosome display are totally in vitro selection methods [Roberts 1999]. They allow a library size as large as 10^{14} different members. The large library size increases the probability to select very rare sequences, and also improves the diversity of the selected sequences. In addition, in vitro selection methods remove unwanted selection pressure, such as poor protein expression, and rapid protein degradation, which may reduce the diversity of the selected sequences. Finally, in vitro selection methods allow the application of in vitro mutagenesis and recombination techniques throughout the selection process. Moreover, although both ribosome display and mRNA display are both in vitro selection methods, mRNA display has some advantage over the former. mRNA display utilizes covalent mRNA-polypeptide complexes linked through puromycin; whereas, ribosome display utilizes stalled, noncovalent ribosome-mRNA-polypeptide complexes and selection stringency is limited. This may cause difficulties in reducing background binding during the selection cycle. Also, the polypeptides under selection in a ribosome display system are attached to an enormous rRNA-protein complex, the ribosome itself, and there might be some unpredictable interaction between the selection target and the ribosome, thus leading to a loss of potential binders during the selection cycle. In contrast, the puromycin DNA spacer linker used in mRNA display technology is much smaller comparing to a ribosome, so having less chance to interact with an immobilized selection target. Thus, mRNA display technology is more likely to give less biased results [Gold 2001].

mRNA display has been used to select high affinity reagents from engineered libraries of linear peptides [Barrick and Roberts 2002; Barrick et al. 2001; Wilson et al. 2001; Baggio et al. 2002], constrained peptides [Baggio et al. 2002], single-domain antibody mimics [Xu et al. 2002], variable heavy domains of antibodies and single-chain antibodies [Chen 2003]. In addition, mRNA-display selections from proteomic libraries have identified cellular polypeptides that bind specific signaling proteins [Hammond et al. 2001] and small-molecule drugs, as well as polypeptide substrates of v-abl kinase [Cujec et al. 2002].

In general, in vitro display technologies prove to be valuable tools for many applications other than merely selecting polypeptide binders. They have great potential for directed evolution of protein stability and affinity, the generation of high-quality libraries by in vitro preselection, the selection of enzymatic activities, and the display of cDNA and random-peptide libraries [Amstutz et al. 2001; Lipovsek and Plückthun 2004].

YEAST DISPLAY

In yeast display (or yeast surface display) a protein of interest is displayed as a fusion to the Aga2p protein on the surface of yeast [Boder and Wittrup 1997, 1998]. The Aga2p protein is naturally used by yeast to mediate cell-cell contacts during yeast cell mating. As such, display of a protein via Aga2p projects the protein away from the cell surface, minimizing potential interactions with other molecules on the yeast cell wall. The use of magnetic separation and flow cytometry in conjunction with a yeast display library is a highly effective method to isolate high affinity protein ligands against nearly any receptor through directed evolution.

Advantages of yeast display over other in vitro evolution methods include eukaryotic expression and processing, quality control mechanisms of the eukaryotic secretory pathway, minimal avidity effects, and quantitative library screening through fluorescent-activated cell sorting (FACS) [Feldhaus and Siegel 2004a, 2004b].

Disadvantages include smaller mutagenic library sizes compared to alternative methods and differential glycosylation in yeast compared to mammalian cells. It should be noted that these disadvantages have not limited the success of yeast display for a number of applications, including engineering the highest monovalent ligand-binding affinity reported to date for an engineered protein [Boder et al. 2000].

BACTERIAL DISPLAY

Similarly to yeast display technology, in bacterial display (or bacterial surface display) libraries of polypeptides displayed on the surface of bacteria can be screened using iterative selection procedures (biopanning), flow or cell sorting techniques [Francisco et al. 1993], thus simplifying the isolation of proteins with high affinity for ligands. Expression of antigens on the surface of non-virulent microorganisms is an attractive approach to the development of high-efficacy recombinant live vaccines [Georgiou et al. 1997]. Finally, cells displaying protein receptors or antibodies are of use for analytical applications and bioseparations.

Chapter 5

PHAGE DISPLAY FOR DIRECTED MOLECULAR EVOLUTION

Phage display is a fundamental tool in protein engineering as well as a method for studying protein-protein, protein-peptide, and protein-DNA interactions that utilizes bacteriophage to connect proteins with the genetic information that encodes them [Smith 1985]. This connection between genotype and phenotype enables large libraries of proteins to be screened and amplified in a process of in vitro selection that imitates the strategy of natural evolution in the test tube. Phage display technology involves the expression of random peptides on the surface of a bacteriophage, displayed as a fusion with one of the viral structural protein. The most common bacteriophages used in phage display are M13 and fd filamentous phage [Smith and Petrenko 1997; Kehoe and Kay 2005], though T4 [Ren and Black 1998], T7 [Rosenberg et al. 1996], and λ phage [Santini et al. 1998] have also been used.

Filamentous phages [Marvin 1998] are flexible, thread-like particles approximately 1 μm long and 6 nm in diameter. The bulk of their tubular capsid consists of 2700 identical subunits of the 50-residue major coat protein pVIII arranged in a helical array with a five-fold rotational axis and a coincident two-fold screw axis with a pitch of 3.2 nm. The major coat protein constitutes 87% of total virion mass. Each pVIII subunit is largely α–helical and rod-shaped; about half of its 50 amino acids are exposed to the solvent, the other half being buried in the capsid. At one tip of the particle the capsid is capped with five copies each of minor coat proteins pVII and pIX; five copies each of minor coat proteins pIII and pVI cap the other end. The capsid encloses a single-stranded DNA. Longer or shorter plus strands —including recombinant genomes with foreign DNA inserts— can be accommodated in a capsid whose length matches the length of the enclosed DNA by including proportionally fewer or more pVIII subunits.

In 1985, recombinant DNA techniques were applied to phage to fashion a new type of molecular chimera that underlies today's phage display technology [Smith 1985]. A foreign coding sequence is spliced in-frame into a phage coat protein gene, so that the "guest" peptide encoded by that sequence is fused to a coat protein, and thereby displayed on the exposed surface of the virion.

In early examples of M13 filamentous phage display, polypeptides were fused to the amino-terminus of either pIII or pVIII in the viral genome [Scott and Smith 1990; Greenwood et al. 1991]. These systems were severely limited because large polypeptides (>10 residues for pVIII display) compromised coat protein function and so could not be efficiently displayed. The development of phagemid display systems solved this problem because, in such systems, polypeptides were fused to an additional coat protein gene encoded by a phagemid vector [Bass et al. 1990]. Multiple cloning sites are sometimes used to ensure that the fragments are inserted in all three possible frames so that the cDNA fragment is translated in the proper frame. The phagemid is then transformed into *Escherichia coli* bacterial cells such as TG1 or XL1-Blue *E. coli*. The phage particles will not be released from the *E. coli* cells until these are infected with helper phage, which enables packaging of the phagemid DNA in a coat composed mainly of wild-type coat proteins from the helper phage but also containing some fusion coat proteins from the phagemid. In phagemid systems, functional polypeptide display has now been demonstrated with all five M13 coat proteins.

By cloning large numbers of DNA sequences into the phage, display libraries are produced with a repertoire of many billions of unique displayed proteins. A phage display library is, in fact, an ensemble of up to about 10 billion recombinant phage clones, each harboring a different foreign coding sequence, and therefore displaying a different guest peptide on the virion surface.

The foreign coding sequence can derive from a natural source, or it can be deliberately designed and synthesized chemically. For instance, phage libraries displaying billions of random peptides can be readily constructed by splicing degenerate synthetic oligonucleotides, obtained by combinatorial approach into the coat protein gene. Displayed peptides can be linear or disulfide-constrained [McLafferty et al. 1993; Ladner 1995; Saggio and Laufer 1993; Luzzago and Felici 1998], aimed to mimic in minute detail similar natural ligands and epitopes.

Recombinant peptides specifically binding a target of interest can be selected from random peptidic libraries (usually from 8- to 20-mer), by a process of affinity selection known as "biopanning".

By immobilizing the target protein to a solid support of some sort (e.g., on the polystyrene surface of an ELISA well or on a magnetic bead), a phage that displays a protein binding to one of those targets on its surface will be captured on the support and remain there while others are removed by washing. Those that remain —generally a minuscule fraction of the initial phage population— can then be eluted in a solution that loosens target-peptide bonds without destroying phage infectivity, propagated simply by infecting fresh bacterial host cells and so produce a phage mixture that is enriched with relevant binding phage and that can serve as input to another round of affinity selection. Phage eluted in the final step (typically after 2-4 rounds of selection) can be used to infect a suitable bacterial host, from which the phagemids can be collected and the relevant DNA sequenced to identify the interacting proteins or protein fragments.

Recent work published by Chasteen *et al* [2006] shows that use of the helper phage can be eliminated by using a novel "bacterial packaging cell line" technology.

In addition, phage selection is not limited to solid support biopanning as described above but has been used also against intact cells for selection of tissue and cell targeting proteins [Samoylova et al. 2003; Spear et al. 2001]. In particular, this technology represents a powerful tool for the selection of peptides binding to specific motifs on whole cells since it is a non-targeted strategy, which also enables the identification of surface structures that may not have been considered as targets or have not yet been identified [Bishop-Hurley et al. 2005].

More recently, in vivo phage display has been used extensively to screen for novel targets of tumor therapy [Schluesener and Xianglin 2004; Li et al. 2006, Lee et al. 2002; Brown C. K. et al. 2000; Zhang et al. 2005].

Phage display is a practical realization of an artificial chemical evolution [Smith and Petrenko 1997]. Using standard recombinant DNA technology, peptides are associated with replicating viral DNAs that include their coding sequences. This confers on them the key properties of evolving organisms: replicability and mutability. Affinity for the target is an artificial analogue to the "fitness" that governs an individual's survival in the next generation. Because selection parameters can be designed and controlled, the phage display is an ideal instrument for directed molecular evolution.

The peptide populations so created are managed by simple microbiological methods that are familiar to all molecular biologists, and they are replicable and therefore nearly cost-free after their initial construction or selection. Therefore, phage display has the overwhelming advantage to be cheap and easy.

The proteins displayed range from short amino acid sequences to fragments of proteins [Wang et al. 1995; vanZonneveld et al. 1995], enzymes [Wang et al.

1996], receptors [Gu et al. 1995; Onda et al. 1995; Sche et al. 1999; Fakok et al. 2000], DNA and RNA binding proteins [Wu et al. 1995; Wolfe et al. 1999; Segal et al. 1999; Isalan and Choo et al. 2000; Cheng et al. 1996] and hormones [Cabibbo et al. 1995; Wrighton et al. 1996; Livnah et al. 1996].

Geysen and his colleagues introduced the term "mimotope" to refer to small peptides that specifically bind a receptor's binding site (and in that sense mimic the epitope on the natural ligand) without matching the natural epitope at the amino acid sequence level [Geysen et al. 1986a, 1986b]; the definition includes cases where the natural ligand is non-proteinaceous. Mimotopes are usually of little value in mapping natural epitopes, but may have other important uses, as identifying new receptors and natural ligands for an "orphan receptor" [Houimel et al. 2002; El-Mousawi et al. 2003], peptides that might act as enzyme inhibitors [Hyde-DeRuyscher et al. 2000; Dennis et al. 2000; Maun et al. 2003; Huang et al. 2003; Lunder et al. 2005a, 2005b; Bratkovic et al. 2005; Nakamura et al. 2001] and receptor agonists or antagonists [Skelton et al. 2002; McConnell et al. 1998; Nakamura et al. 2002; Hessling et al. 2003], "epitope discovery" [Wang and Yu 2004; Petit et al. 2003; Leinonen et al. 2002; Coley et al. 2001; Myers et al. 2000; Rowley et al. 2000], design of DNA-binding proteins [Wu et al. 1995; Wolfe et al. 1999; Segal et al. 1999; Isalan and Choo et al. 2000; Cheng et al. 1996], source of new materials [Smith and Petrenko 1997; Souza et al. 2006; Nam et al. 2004]. The proteins so synthesized can indeed be considered as non-extant, and artificial proteins with random sequence can be displayed [Nakashima et al. 2000], which permits the terminology of "never born proteins" (NBPs).

Phage display is an exponentially growing research area, and numerous reviews covering different aspects of it have been published [Felici et al. 1995; Cortese et al. 1995; Smith and Petrenko 1997]. In conclusion, then, it is a useful tool in protein engineering as well as in functional genomics and proteomics, in drug discovery and, we can say, in synthetic biology.

Chapter 6

ANTIBODY PHAGE

Over the past decade, phage-displayed antibody fragments have been the subject of intensive research [reviewed in Dall'Acqua and Carter 1998; Griffiths and Duncan 1998]. As a result, antibody phage libraries have become practical tools for drug discovery and several phage-derived antibodies are in advanced clinical trials. Phage display has provided approximately 30% of all human antibodies now in clinical development [Kretzschmar and von Ruden 2002].

Large collection of antibody fragments have been displayed on phage particles, and successfully screened with different antigens [Hust and Dubel 2004]. Since the first demonstration that it was possible to display functional antibody fragments on the surface of filamentous phage [McCafferty et al. 1990; Hoogenboom et al. 1991; Barbas et al. 1991; Breitling et al. 1991; Garrard et al. 1991], the development of this technique has led to the construction of recombinant antibody libraries displayed on the bacteriophage surface. The selection of antibodies by phage display basically relies on several factors: first, the ability to isolate or synthesize antibody gene pools to construct large, highly diverse libraries; second, the possibility to express functional antibody fragments in the periplasmic space of *E. coli* [Better et al. 1988; Skerra and Plückthun 1988]; and third, the efficient coupling of expression and display of the antibody protein with the antibody's genetic information being packaged in the *E. coli* bacteriophage [Smith 1985; McCafferty et al. 1990].

Filamentous bacteriophage such as M13 and its coat protein pIII are most often used for antibody display, although T7 bacteriophage also reportedly allow antibody display [Rosenberg et al. 1996].

Both scFv (single chain Fv fragments, where the heavy and light chain V regions are linked by a linker peptide) [McCafferty et al. 1990] and Fab

(Fragment antigen binding dimers) [Hoogenboom et al. 1991; Garrard et al. 1991; Chang et al. 1991] formats have been used successfully in antibody libraries displayed on phage, with the former representing the more popular choice [Carmen and Jermutus 2002]. Large repertoires of heavy and light chain V regions can be obtained through amplification by the polymerase chain reaction from the B cells of an immunized animal (usually extracted from the spleen) [Clackson et al. 1991], or hybridoma cells generated from such an animal [Orlandi et al. 1989; Chiang et al. 1989], or even immunized humans ("immunized libraries") [Persson et al. 1991; Burton et al. 1991; Graus et al. 1997; Cai and Garen 1995]; these repertoires will contain antibodies that are biased towards the immunogen, based on the host's immune response.

Alternative approaches are constituted by the "semi-synthetic libraries", where germ-line heavy and light chain V regions, cloned from human B cells, are assembled and synthetic randomization is used to introduce additional diversity at the CDR3 region to increase the repertoire [Barbas et al. 1992], and the "naïve libraries" heavy and light chain variable regions are amplified from the naïve Ig repertoire of a healthy human donor and randomly combined to produce the phage-displayed library [Carmen and Jermutus 2002; Marks et al. 1991].

An important advance has been the development of high-quality libraries with completely synthetic complementarity-determining regions. Knappik *et al* [2000] have constructed a library in which a limited number of human frameworks are used and diversity is introduced by means of synthetic cassettes. Such a system is very amenable to the generation of therapeutic antibodies because preferred frameworks can be used and affinity maturation is aided by the use of defined mutagenic cassettes.

The construction of large, high-quality antibody phage libraries has also been aided by the introduction of improved in vivo recombination systems [Sblattero and Bradbury 2000].

The selection of antibodies from phage libraries consists of two main steps: panning and screening. During panning, library phage preparations are incubated with the antigen of choice, unbound phage are discarded and remaining phage recovered after several washing steps by disrupting the phage–antigen interaction without compromising the phage infectivity (e.g. by applying pH-gradients, competitive elution conditions or proteolytic reactions). Recovered phage subsequently are amplified by infecting *E. coli* and further round(s) of panning are applied, yielding a polyclonal mixture of phage antibodies enriched for antigen-specific binders. The purified antigen can be attached to a solid support (to plastic, by adsorption, or to beads or a column matrix), the phage library run over the support, and antigen-bound phage retrieved after rinsing the support by elution;

alternatively, the binding event can take place with antigen in solution, for example, using biotinylated antigen or unlabelled antigen, and the antigen-bound phage might be retrieved by incubation with streptavidin-coated magnetic beads or other ligands that capture the antigen.

The in vitro selection procedure can be performed for function, besides for binding capabilities. Selections can be performed under conditions that mediate the selection of phage antibodies with a particular characteristic, for example, under reducing conditions to retrieve disulfide-free yet stable antibodies [Proba and Wörn 1998], or in the presence of proteases to select for well-folded molecules [Kristensen and Winter 1998]. Antibodies might also be selected with or for a particular functional activity, for example, for receptor cross-linking, signalling, gene transfer or catalysis [Hoogenboom et al. 1998].

The screening process involves subsequently converting the polyclonal mixture obtained by panning into monoclonal antibodies. To this end, *E. coli* cells are infected with the phage pool, plated on selective agar dishes, and single colonies are picked. Thus, highly specific, monoclonal antibody clones are obtained, from which the antibody genes can be readily isolated for further analyses and/or engineering [Rader and Barbas 1997; Griffiths and Duncan 1998; Hoogenboom and Chames 2000; Siegel 2002].

A single phage antibody library can be distributed to thousands of users and serve as the source of cloned antibodies against an unlimited array of antigens. Because selection is based solely on affinity, many toxic and biological threat agents that could not be used to immunize animals without their prior inactivation can nevertheless serve as "native antigens" in this artificial immune system. Furthermore, phage display allows selecting of antibodies recognizing unique epitopes on biological agents that may be missed in hybridoma screening [Emanuel et al. 1996]. Another advantage of phage display contrasting it to the hybridoma technique is that the quantity of antigens required for selection of phage antibodies may be surprisingly small [Liu and Marks 2000], and the properties of selected probes can be further improved by affinity maturation and molecular evolution [Chowdhury 2002; Deng et al.1994; Worn and Plückthun 2001]. Thus, for many purposes, this system may well come to replace natural immunity in animals [Liu and Marks 2000]. With phage display, as in the in vivo immune system, antibodies can be affinity-matured in a stepwise fashion, by incorporating mutations and selecting variant cells with decreasing amounts of antigen [reviewed in Hoogenboom 1997]. Various procedures for introducing diversity in the antibody genes have been described, ranging from more-or-less random strategies (e.g. V-gene chain shuffling, error-prone PCR, mutator strains or DNA shuffling), to strategies targeting the CDR regions of the antibody for mutagenesis (e.g. oligonucleotide-directed mutagenesis, PCR). One possible disadvantage of this in vitro process is that the affinity improvement can be accompanied by the appearance of a modified fine-specificity or unwanted cross-reactivity [Ohlin et al. 1996], which the natural immune system might quickly remove. Extensive screening of the in vitro affinity-matured antibodies for changes in specificity is thus required.

There are several examples where phage antibodies selected against various biological threats have been used beneficially in various detection platforms [reviewed in Iqbal et al. 2000] and for therapeutic applications [reviewed in Kretzschmar and von Ruden 2002].

Chapter 7

PHAGE AS BIOSELECTIVE PROBES

Development of systems for monitoring the environment and food for biological threat agents is a challenge because it requires environmentally stable, long lasting, sensitive and specific diagnostic probes capable of tight selective binding of pathogens in unfavorable conditions. To respond to the challenge, large financial investments and extensive collaborative efforts of specialists in different areas of science and technology are required. In the last years, probe technology is being revolutionized by utilizing methods of combinatorial chemistry and directed molecular evolution. In particular, phage display is recently identified [Brigati et al. 2004; Petrenko and Smith 2000; Petrenko and Sorokulova 2004; Petrenko and Vodyanoy 2003] as a new technique for development of diagnostic probes which may meet the strong criteria —fastness, sensitiveness, accuracy, and inexpensiveness— for biological monitoring [Al-Khaldi et al. 2008, 2009].

The most commonly used bioselective probes are antibodies, but a variety of other bio-organic molecules have also been effectively used as for example peptides, enzymes, lectins, carbohydrates, nucleic acids, aptamers, recombinant proteins or molecularly imprinted polymers, and more recently whole cells. However, none of these types of recognition interfaces could meet detection performance requirements completely. Recombinant phage provide another source of high quality detection reagents. A number of probes have been selected from phage display libraries with high specificity and selectivity for a wide range of targets [Kouzmitcheva et al. 2001; Petrenko et al. 1996; Petrenko and Smith 2000; Romanov et al. 2001; Samoylova et al. 2003], including small molecules [Saggio and Laufer 1993], receptors [Balass et al. 1993; Legendre and Fastrez 2002], virus [Gough et al. 1999], bacterial spores [Knurr et al. 2003; Steichen et al. 2003; Turnbough 2003; Brigati et al. 2004], and whole-cell epitopes [Carnazza et al. 2007, 2008; Cwirla et al. 1990; Olsen et al. 2006; Petrenko and Sorokulova 2004; Sorokulova et al. 2005; Stratmann et al. 2002; Yu and Smith 1996].

Affinity-selected landscape phage probes for *Salmonella typhimurium* were demonstrated to possess the specificity, selectivity and affinity of monoclonal antibodies, and used as probes for the detection of *S. typhimurium* [Sorokulova et al. 2005].

The performance of the probes in the detection of biological agents was demonstrated using a quartz crystal microbalance (QCM) [Petrenko and Vodyanoy 2003; Yang et al. 2008], acoustic wave piezoelectric transducers [Olsen et al. 2006, 2007], electrochemical sensors [Yang et al. 2006], wireless magnetoelastic biosensors and microcantilevers [Wan et al. 2007a; Johnson et al. 2008; Fu et al. 2007], and PSR sensors [Nanduri et al. 2007a] in which phage immobilized on a gold electrode reacted with their analytes in solution-phase [Petrenko 2008b].

At the same time, phage antibody chip strategy proved its efficacy in preliminary diagnostics of cancer or other diseases applications [Hong et al. 2004]; phage arrays were constructed by using five clones, displaying respectively four scFv from mouse and one humanized scFv. The targets were Cy3 fluorescence labeled protein extracts from normal lymphocytes and tumorous HeLa cells. Fluorescence intensity of phage was exploited to indicate overexpression of some proteins in the tumor cell line when compared to normal lymphocytes.

Another similar proof-of-principle experiment for phage antibody chips was also reported from the same research group [Bi et al. 2007]. A protein microarray spotted directly with ninety-six phage-displayed antibody clones, half of them deriving from cell panning with leukocytes from healthy donors, and half from panning with acute myeloid leukemia leukocytes, was created to discriminate between recognition profiles of samples from healthy donors and leukemia patients. The signals of nine of those probes showed significant difference between normal and leukemia samples.

In general, fusion of a peptide to the pIII minor coat protein, located on the tip of the phage capsid, is probably not optimal for obtaining phage probes because this expression format does not take advantage of the avidity effect gained when the binding peptides are displayed multi-valently on the major coat protein pVIII [Liao et al. 2005]. This is because pIII is the minor protein of wild phage and there are only five copies of pIII on the tip of the phage. In the pIII display system, scFv is always expressed mono-valently, and most probably scFv will be either orientated parallel to or in contact with the surface which may restrict the freedom of scFv recognition. On the contrary, the result is surprising in pVIII display system, with a ratio of the positive signal to the negative of 3000:1. This is attributed to the amount and status of pVIII of phage, which forms the tube of phage with about 2700 copies. Using pVIII display system, not only the number of fused scFv is increased, but also the orientation is improved because there is always half of the displayed scFv stretching freely out into solution. Thus, phage antibody chip by pVIII display system seems to be very promising.

The overall data strongly suggest that new generation of selective and inexpensive phage-derived probes will serve as efficient substitutes for antibodies in separation, concentration and detection systems employed for clinical and environmental monitoring, for example by developing rapid diagnostic arrays.

The main advantages of phage probes include: the simplicity of manipulation of the phage libraries, their great variability, high binding affinity, great stability and the possibility to select probes to targets of different nature, also to small molecules or toxic compounds or immunosupressants against which it is difficult to raise natural antibodies.

In fact, while sensitive and selective, antibodies have numerous disadvantages for use as diagnostic biodetectors in biological monitoring, including high cost of production, low availability, great susceptibility to environmental conditions [Shone et al. 1985] and the need for laborious immobilization methods to sensor substrates [Petrenko and Vodyanoy 2003]. The phage probes affinity-selected from random libraries for specific and selective binding to biological targets can be an effective alternative to antibodies [Sorokulova et al. 2005].

They can act as antibody surrogates, possessing distinct advantages including durability, stability, standardization and low-cost production, while achieving equivalent specificity and sensitivity [Petrenko et al. 1996; Petrenko and Vodyanoy 2003]. A selected phage itself can be used as a probe in a detection device, without chemical synthesis of the displayed peptide or fusion to a carrier protein. For example, to be used in an automated fluorescence based sensing assay [Goldman et al. 2000] or FACS [Turnbough 2003], phage, exposing thousands of reactive amino-groups, can be conjugated with fluorescent labels and, in this format, successfully compete with antibody-derived probes.

The use of antibodies as diagnostic probes outside of a laboratory may be problematic because they are often unstable in severe environmental conditions. Environmental monitoring requires stable probes, such as landscape phage, that carry thousands of foreign peptides on their surfaces, are as specific and selective as antibodies, and can operate in non-controlled conditions. Filamentous phage are probably the most stable natural nucleoproteins capable of withstanding high temperatures (up to 80°C), denaturing agents (6–8 mol/L urea), organic solvents [e.g., 50% alcohol or acetonitrile], mild acids (pH 2), and alkaline solutions. The thermostability of a landscape phage probe was recently examined in comparison with a monoclonal antibody specific for the same target [Brigati and Petrenko 2005]. They were both stable for greater than six months at room temperature, but at higher temperatures the antibody degraded more rapidly than the phage probe. Phage retained detectable binding ability for more than 6 weeks at 63°C, and 3 days at 76°C. Similarly, phage-coated magnetoelastic biosensors preserved about 49, 40, and 25% of their binding activity after three months of storage at 25, 45 and 65°C, respectively [Wan et al. 2007b], whereas the antibody-coated sensors showed no binding affinity after only five days at 65 and 45°C. These results confirm that phage probes are highly thermostable and can function even after exposure to high temperatures during shipping, storage and operation.

Phage-derived probes inherit the extreme robustness of the wild-type phage and, in addition, allow fabrication of bioselective materials by self-assemblage of phage or their composites on metal, mineral, or plastic surfaces [Petrenko and Vodyanoy 2003]. The recombinant phage probes appear to be highly amenable to simple immobilization through physical adsorption directly to the sensor surface, thus providing another engineering advantage while maintaining biological functionality [Carnazza et al. 2007, 2008; Olsen et al. 2006; Sorokulova et al. 2005]. This property allows phage to be used as a recognition element in biosensors, like an inexpensive standard construction material that allows fabrication of bioselective layers by self-assembly of virions or their composites on solid surfaces [Nanduri et al. 2007c].

Protocols for immobilizing bacteriophage particles on solid surfaces have been described since the inception of phage display technology [Smith 1992]. Purified phage particles can be either directly coated to plastic surfaces, or anti-phage monoclonal antibodies can be used to tether phage particles onto the surface of multiwell plates directly from crude supernatants of infected bacteria, without any previous purification step [Dente et al. 1994].

With respect to methods relevant for phage probes used in protein chip and biosensor applications, many different technical approaches of immobilization have been exploited, such as physical adsorption [Wan et al. 2007a,b; Olsen et al. 2006; Nanduri et al. 2007a-c; Lakshmanan et al. 2007], covalent binding [Yang et al. 2008], and molecular recognition [Petrenko and Vodyanoy 2003; Olsen et al. 2007]. For examples, phage has been immobilized by direct physical adsorption to the gold surface [Nanduri et al. 2007c], by peptide bond between amino residue on phage and carboxyl terminal on surface [Ploug et al 2001], by disulfide bond between one thiol group on phage and another on surface [Dultsev et al. 2001], and by specific recognition between hexahistidine tag on phage and nickel coated surface [Finucane et al. 1999].

In addition, phage probes can functionalize surface with less steric hindrance than antibodies, thus allowing a higher binding avidity for the target per surface unit. In fact, on the same surface unit, a greater number of phage and with a more correct orientation can be patterned in comparison to antibodies.

Therefore, different phage clones could be selected specifically binding to isolated proteins, enzyme or inorganic material, as well as to different microbial species, thus with a single array different targets might be detected at once, by performing in parallel several different assays in real-time, within the same miniaturized substrate. In this way, standardizing data from multiple separate experiments will be unnecessary, and truly meaningful comparisons may be derived. This could ultimately translate to a much lower cost per test. Much of the promise of these microarrays relies in their small dimensions, which reduce sample and reagent requirements and reaction times, while increasing the amount of data available from a single assay.

Furthermore, phage probes may find application as bio-recognition elements of real-time biosensor devices. Recombinant phage selected by phage-display selectively recognize and specifically bind complex target structures, such as bacterial cells, thus they can be used to develop rapid diagnostic arrays. In fact, traditional diagnostic systems usually involve a multi-step detection method with the use of labeled secondary antibodies. Phage-displayed detection microsystems could be considered one-step, simultaneously bind and identify the target microorganism, with no need of further characterization steps.

On the other hand, in nanobiotechnology, acquisition of abundant probes and label-free, high sensitive detection now become the important issues. Generally, labeling tends to cause the deactivation of protein, owing to protein complex three-dimensional structure. The combination of phage-displayed probes and the label-free, real-time detection method based on surface plasmon resonance (SPR) technique has been proved to be fit for proteomics research. Lytic phage were used as probes on an SPR platform, SPREETATM (Texas Instrument, US), for detection of *Staphylococcus aureus* [Balasubramanian et al. 2007], and scFv antibody displayed on Lm P4:A8 phage pIII protein was used to detect *Listeria monocytogenes* and its virulence factor ActA [Nanduri et al. 2007b]. More recently, phage display technique and SPR detection method were combined to acquire abundant specific capture molecules and realize a label-free and high-sensitive protein chip [Liu et al. 2008]. A 12-amino acid peptide displayed on phage M13 coat protein pIII was selected as the probe, and it was immobilized on 11-mercapto-undecanoic acid sensor chip to fabricate a reusable phage-displayed protein chip. The interaction between the peptide and the specific ligand protein was detected on the BIAcore3000 (BIAcore AB, Sweden). Experimental results showed that the phage-displayed protein chip can act as a useful tool in proteomics research.

On the whole, this would allow the use of phage probes in development of analytical devices for detecting and monitoring agents under any conditions that warrant their recognition, including clinical based diagnostics and biological warfare applications, spanning several potential markets including biomedical and industrial use, monitoring and proteomics research.

Chapter 8

PHAGE-DERIVED NANOMATERIALS

In the recent years, phage display evolved into the discipline of material science, presenting phage not only as an instrument for peptide and antibody discovery, but also as a prospective nanomaterial that can be easily tailored using routine genetic engineering manipulations. This merge of phage display technologies with nanotechnology appears promising for applications in different fields, such as medical diagnostics and monitoring, molecular imaging, targeted therapy and gene delivery, vaccine development, as well as bone and tissue repair [Petrenko 2008a].

Viruses are highly organized supramolecular arrays put together by a combination of non-covalent self-assembly and genetic programming. One of the attractions of viruses as nanostructured materials is that their surface chemistry is highly amenable to fine, site-specific and inheritable modification: the proteins that constitute the coat can be altered by introducing the appropriate sequence into the gene that encodes them.

Viral capsids offer the advantages of being robust and monodisperse, and can exhibit various sizes and shapes. Filamentous bacteriophage [Wilson and Finlay 1998] is a flexible rod, of 6 nm diameter and 800-2000 nm length, depending on the genome length. The capsid is mainly constituted by a protein (pVIII) arranged in a helical array with a 5-fold symmetry axis around a single-stranded DNA molecule. A series of aspartate and glutamate residues ensure a negative potential on the surface, and one tryptophan residue is buried in the hydrophobic region responsible for packing of the capsid.

In a nanobiotechnological approach, fusion phage serves not only as a growth-supporting and peptide-exposing carrier, but operates as a nanotube decorated by thousands of foreign peptides, whose composition determines the integrated physico-chemical characteristics of the whole nanoparticle. In this concept, virus capsids serve as versatile scaffolds that can be used for producing nanomaterials. Landscape phage obtained by phage-display technology might be looked on as a new kind of submicroscopic "fiber" [Smith and Petrenko 1997]. Each phage clone is a type of fiber with unique surface properties. These fibers are not synthesized one by one with some use in mind. Instead, billions of fibers are constructed, propagated all at once in a single vessel and portions of this enormous population are distributed to multiple end-users with many different goals. Each user must devise a method of selecting from this population those fibers that might be suitable for his or her particular applications by affinity selection or whatever other selection principle can be conceived. The phage-display approach provides a physical linkage between the peptide-substrate interaction and the DNA that encodes that interaction. Localizable or global emergent properties cannot be transferred from the virion surface to another medium; any application that depends on such properties must therefore use phage themselves as the new material.

Long rod shaped M13 viruses were used to fabricate one dimensional micro- and nano-sized fibers by mimic the spinning process of the silk spider [Belcher and Lee 2008]. After blending with highly water soluble polymer, polyvinyl 2-pyrolidone (PVP), M13 viruses were spun into continuous uniform virus blended PVP fibers that showed intact infecting ability to bacterial hosts after suspending in the buffer solution.

Given that filamentous phage are resistant to harsh conditions such as high salt concentration, acidic pH, chaotropic agents, and prolonged storage, they are suitable candidate building blocks to meet the challenges of bottom-up nano-fabrication.

The first phage-derived nanomaterial to be used as an artificial specific immunogen and a potential vaccine was proposed by the research group of Petrenko and co-workers [Minenkova et al. 1993], which obtained a pVIII-fusion phage displaying epitopes of p17 Gag protein of the HIV1 virus.

Moreover, the pIII minor capsid protein of the phage can be easily engineered genetically to display ligand peptides that will bind to and modify the behaviour of target cells in selected tissues. Thus, the tactic of integrating phage display technology with tailored nanoparticle assembly processes offers opportunities for reaching specific nano-engineering and biomedical goals [Giordano et al. 2001; Trepel et al. 2002; Arap et al. 2002; Langer and Tirrell 2004; LaVan et al. 2002].

Recent efforts focused on the bottom-up assembly of functional nanosystems from nanoscale building blocks have led to substantial advances. M13 viral particles with distinct substrate-specific peptides expressed on the filamentous capsid and the ends were demonstrated to provide a generic template for programmable assembly of complex nanostructures. Phage clones with gold-binding motifs on the capsid and streptavidin-binding motifs at one end were created and used to assemble Au and CdSe nanocrytals into ordered one-dimensional arrays and more complex geometries [Huang et al. 2005]. Initial studies show such nanoparticle arrays can further function as templates to nucleate highly conductive nanowires that are important for addressing/interconnecting individual nanostructures.

The steric constraints inherent to the competitive charge binding between M13 viruses and two oppositely charged weak polyelectrolytes leads to inter-diffusion and the virtual "floating" of viruses to the surface, resulting in the spontaneous formation of a two-dimensional monolayer structure of viruses on a cohesive polyelectrolyte multilayer [Yoo et al. 2006]. This viral-assembled monolayer can be a biologically tunable scaffold to nucleate, grow and align nanoparticles or nanowires over multiple length scales, thus representing an interface that provides a general platform for the systematic incorporation and assembly of organic, biological and inorganic materials.

Another approach for fabrication of spontaneous, biologically active molecular networks consisting of phage directly assembled with gold (Au) nanoparticles has been reported [Souza et al. 2006]. In this work, it was shown that such networks are biocompatible and preserve the cell-targeting and internalization attributes mediated by a displayed peptide and that spontaneous organization (without genetic manipulation of the pVIII major capsid protein), and optical properties can be manipulated by changing assembly conditions. By taking advantage of Au optical properties, Au–phage networks were generated that, in addition to targeting cells, could function as signal reporters for fluorescence and dark-field microscopy and near-infrared (NIR) surface-enhanced Raman scattering (SERS) spectroscopy. Notably, this strategy maintains the low-cost, high-yield production of complex polymer units (phage) in host bacteria and bypasses many of the challenges in developing cell-peptide detection tools, such as complex synthesis and coupling chemistry, poor solubility of peptides, the presence of organic solvents, and weak detection signals. These networks can effectively integrate the unique signal reporting properties of Au nanoparticles while preserving the biological properties of phage. Together, the physical and biological features within these targeted networks offer convenient multifunctional integration within a single entity with potential for nanotechnology-based biomedical applications.

Furthermore, the bacteriophage MS2 was used as a potential vector for transporting the anti-tumour drug taxol, by linking it covalently to the acid-labile chemical linker groups attached to the inside of the spherical virion [Kooker et al. 2004].

Of particular value would be methods that could be applied to materials with interesting electronic or optical properties. Organizing ordered inorganic nanoparticles by using biological structures as templates is essential for constructing nano-devices with new functionalities. Nature shows how soluble molecules capable of recognizing and binding to specific materials can be used to shape and control the growth of crystals and other nanostructures. There is no need to rely on the complexity of the immune system in order to conduct combinatorial searches for new peptides of this sort. Although natural evolution has not selected for interactions between biomolecules and such materials, phage-display libraries can be successfully used to identify, develop and amplify binding between organic peptide sequences and inorganic metal and semiconductor substrates.

Peptides with limited selectivity for binding to metal surfaces and metal oxide surfaces have been successfully selected [Brown 1992; Brown 1997]; other researchers have used phage display to select peptides against synthetic polymers such as polystyrene [Adey et al. 1995] and yohimbine-imprinted methacrylate polymer for molecular-imprinted receptors [Berglund et al. 1998].

This approach was then extended and it was shown that combinatorial phage-display libraries can be used to evolve peptides that bind to a range of semiconductor surfaces with high specificity, depending on the crystallographic orientation and composition of the materials used [Whaley et al. 2000].

Phage-display libraries, based on a combinatorial library of random peptides-each containing 12 amino acids-fused to the pIII coat protein of M13, provided 10^9 different peptides that were reacted with crystalline semiconductor structures. Five copies of the pIII coat protein are located on one end of the phage particle, accounting for 10-16 nm of the particle. The experiments utilized different single-crystal semiconductors for a systematic evaluation of the peptide-substrate interactions. In this way, 12-mer peptides could be identified that bind to specific crystal faces of GaAs, as well as to the surfaces of GaN, ZnS, CdS, Fe_3O_4 and $CaCO_3$. Peptide binding selective for the crystal composition (for example, binding to GaAs but not to Si) and crystalline face (for example, binding to GaAs(100), but not to GaAs(111)B) was demonstrated. In addition the preferential attachment of phage to a zinc-blended surface in close proximity to a surface of differing chemical and structural composition was reported, demonstrating the high degree of binding specificity for chemical composition. These recognition **peptides might provide selective "glues" for assembling** inorganic nanocrystals into complex arrangements, or for attaching them to other biomolecules for labelling or transport.

Subsequently, phage display has been used again in selecting unique peptides against inorganic semiconductor materials [Flynn et al. 2003; Sano and Shiba 2003].

Brown's work on polypeptides that will bind to specific metals [Brown 1997] has been extended by Sarikaya and coworkers [2003] to make so-called GEPIs (genetically engineered polypeptides for inorganics) that bind a host of materials.

Again, the peptides are prepared by combinatorial shuffling of sequences, coupled to a phage-display screening process. Some of these GEPIs exhibit the ability to modify crystal growth, for example switching the morphology of gold nanocrystals from cubo-octahedral (the equilibrium form) to flat triangular or pseudo-hexagonal forms [Brown S. et al. 2000].

In addition, M13 bacteriophage surfaces have been engineered with recognition peptides [Whaley et al. 2000] so that they bound ZnS or CdS, acting as templates for the synthesis of polynanocrystalline nanowires [Mao et al. 2003, 2004]. Lee *et al* [Lee et al. 2002] combined recognition peptides with the self-organizing property of rod-shaped M13 bacteriophage to arrange inorganic nanocrystals into an ordered superstructure. The viruses spontaneously packed in concentrated solution into a layered, tilted liquid-crystalline phase. When their coats were tipped with a peptide 9-mer that bound to ZnS, the viruses act as "handles" for arranging ZnS nanocrystals into composite layers with a roughly 700 nm periodicity.

Similar modifications were made to the tubular protein sheath of the tobacco mosaic virus (TMV) so that it can bind metal ions such as cobalt, potentially enabling the virus to template magnetic nanowires and nanotubes [Schlick et al. 2005]. Each TMV tube is 300 nm long, made up of 2100 identical protein subunits. Interestingly, the wedge-shaped proteins were also able to form a variety of other potential template structures, such as shorter tubes and disks, depending on parameters such as pH and ionic strength. Functionalization of these structures with chromophores could provide mimics of the disk-shaped light harvesting complexes of photosynthetic bacteria.

More recently, a new memory effect function was reported in the hybrid system composed of TMV conjugated with platinum nanoparticles (TMV–Pt) [Tseng et al. 2006]. The augmentation of electrical conductivity in this TMV–Pt nanocomposite modifies its properties and makes it a suitable candidate for electronic applications. The function of the TMV is not just to provide a backbone for the organization of discrete nanoparticles (NP). Indeed, the TMV consists of an RNA core with rich aromatic rings, such as guanine, which can behave as charge donors, and of coat proteins, which separates the RNA and Pt NP and act as the energy barrier. These interactions between the RNA and proteins in the TMV with the Pt-NPs are responsible for charge trapping for data storage and tunnelling process in high conductance state, thus creating a conductance switching behaviour and a repeatable memory effect in the TMV–Pt devices.

On the other hand, ring-shaped viruses from genetically modified M13 with two different binding peptides at each end were created [Nam et al. 2004]. When a bifunctional linker molecule binding to the two peptides was added, it secured the flexible rod-shaped viruses into rings about 200 nm in diameter. These were proposed to be used as templates for making nanoscale magnetic rings for magnetic data storage [Ball 2005].

Reviews [Sarikaya et al. 2003; Sarikaya et al. 2004] have highlighted the application of phage display in selecting peptides to functionalize biomaterials such as titanium. More recently, a unique strategy for surface functionalization of an electrically conductive polymer, chlorine-doped polypyrrole (PPyCl), which has been widely researched for various electronic and biomedical applications, has been developed [Sanghvi et al. 2005]. A M13 bacteriophage library was used to screen 10^9 different 12-mer peptide inserts against PPyCl, a binding phage was isolated, and the stability and specificity, strength and mechanism of its binding to PPyCl were assessed. In these studies, phage display was used to select for peptides that specifically bound to an existing biomaterial, PPy, and were subsequently used to modify the surface of PPy. PPy is a conductive synthetic polymer that has numerous applications in fields such as drug delivery [Konturri et al. 1998] and nerve regeneration [Schmidt 1997; Valentini et al. 1992], and has been used in biosensors and coatings for neural probes [Vidal et al. 1999; Cui et al. 2001]. Different dopant ions such as chloride, perchlorate, iodine and poly(styrene sulphonate) can be used during electrochemical synthesis to provide the material with varying properties (for example, conductivity, film thickness and surface topography). PPyCl does not contain a functional group for biomolecule immobilization, making it a suitable model polymer for functionalization using a peptide selected with phage display. Further, the specific peptide selectively binding PPyCl was joined to a cell adhesive sequence and used to promote cell attachment on PPyCl, to serve as a bi-functional linker. The use of the selected peptide for PPyCl by phage display can be extended to encompass a variety of therapies and devices such as PPy-based drug delivery vehicles [Konturri et al. 1998], nerve guidance channel conduits [Schmidt et al. 1997; Valentini et al. 1992], and coatings for neural probes [Cui et al. 2001]. Furthermore, this strategy for surface functionalization can be extended to immobilize a variety of molecules to PPyCl for numerous other applications. In addition, phage display can be applied to other existing polymers (including those that are already approved and/or those polymers that lack functional chemical groups for coupling reactions like PPyCl) to develop bioactive hybrid-materials without altering their bulk properties.

Selection of peptides using phage display thus represents a simple and versatile alternative to methods based on electrostatic and hydrophobic interactions between two moieties to achieve functionalization of surfaces. It is theoretically possible to design bivalent recombinant phage with two-component recognition: such phage have the potential to bind to specific locations on a semiconductor structure by peptides displayed on pIII protein and simultaneously to specific target (molecules or cells to be captured) by peptides displayed on pVIII coat protein.

Apart from the relevance of phage template to direct orientation and nucleation of nanomaterials, it found also a novel application in the context of single-molecule (SM) biophysics, which remained heavily reliant on a few accessible assays, such as the gliding filament, the tethered bacterium, and the tethered bead [Block 1997]. Filamentous bacteriophage capsids have the dual responsibility of safeguarding viral DNA as well as locating and binding host receptors during infection; as a result, they are robust, yet adaptable, structures. Genetically engineered constructs can display different reactive species at each of the filament ends and along the major capsid, and the resulting hetero-functional particles have been shown to consistently tether microscopic beads in solution, becoming a strong and versatile biopolymer alternative to dsDNA in constructing the instrumental tethered bead assay [Khalil et al. 2007]. The M13 system appears well suited for higher force studies, such as for protein extension and distortion, particularly now that the force capability of optical tweezers is increasing. Furthermore, the M13 template can provide multiple, localized attachments to the species of interest to prevent or delay detachment.

One of the encouraging messages emerging from such efforts to use essentially biological structures for nanotechnology is that the potential hurdle of **interfacing seems not to be a problem: that is to say, biology is clearly "plastic"** enough to accommodate unfamiliar materials from the inorganic world. Overall, this can be regarded as a kind of synthetic biology in that it involves the reshaping and redirecting of natural molecular systems, typically using the tools of protein and genetic engineering.

Chapter 9

PHAGE PERSPECTIVES IN NANOBIOTECHNOLOGY

In recent years it has been recognized that recombinant bacteriophages have several potential applications in the modern biotechnology industry: they have been proposed, beside for the above described sophisticated design of antibody drugs, detection of pathogenic bacteria and new biomaterials, as alternatives to antibiotics (phage-therapy); as delivery vehicles for protein and DNA vaccines; as gene therapy delivery vehicles; and as tools for screening libraries of proteins, peptides or antibodies. This diversity and the ease of their manipulation and production mean that they have potential uses in research, therapeutics and manufacturing in both the biotechnology and medical fields [for a review, Clark and March 2006].

Phage-display libraries can be screened in several ways to isolate displayed peptides or proteins with practical applications [Benhar 2001; Willats 2002; Wang and Yu 2004]. For example, it is possible to isolate displayed peptides binding target proteins with affinities similar to those of antibodies, which can then be used as therapeutics that act either as agonists or through the inhibition of receptor–ligand interactions [Ladner et al. 2004].

Furthermore, phage-displayed peptides may be used as signal peptides able to trigger complex cell responses. Overall, phage-display selection of peptides mimicking ligands of cell receptors involved in modulating cell processes such as proliferation, apoptosis and differentiation, opens the door for their potential applications, respectively, in regenerative medicine, anti-tumoral development and stem cell differentiation.

VACCINE DELIVERY

Phage have been used as potential vaccine delivery vehicles in two different ways: by directly vaccinating with phage carrying vaccine antigens on their surface or by using the phage particle to deliver a DNA vaccine expression cassette that has been incorporated into the phage genome [Clark and March 2004a, 2004b]. In phage-display vaccination, phage can be designed to display a specific antigenic peptide or protein on their surface. Alternatively, phage displaying peptide libraries can be screened with a specific antiserum to isolate novel protective antigens or mimotopes –peptides that mimic the secondary structure and antigenic properties of a protective carbohydrate, protein or lipid, despite having a different primary structure [Folgori et al. 1994; Phalipon et al. 1997]. The serum of convalescents can also be used to screen phage-display libraries to identify potential vaccines against a specific disease, without prior knowledge of protective antigens [Meola et al. 1995]. Rather than generating a transcriptional fusion to a coat protein, substances can also be artificially conjugated to the surface of phage after growth, which further increases the range of antigens that can be displayed [Molenaar et al. 2002].

More recently, it has also been shown that unmodified phage can be used to deliver DNA vaccines more efficiently than standard plasmid DNA vaccination [Clark and March 2004a, 2004b; March et al. 2004; Jepson and March 2004]. The vaccine gene, under the control of a eukaryotic expression cassette, is cloned into a standard lambda bacteriophage, and purified whole phage particles are injected into the host. The phage coat protects the DNA from degradation and, because it is a virus-like particle, it should target the vaccine to the antigen presenting cells.

TARGETED THERAPY

One particularly novel use for phage-displayed peptides is in targeted therapy. One example was in the development of a nasally delivered treatment against cocaine addiction [Dickerson et al. 2005]: whole phage particles delivered nasally can enter the central nervous system where a specific phage-displayed antibody can bind to cocaine molecules and prevent their action on the brain.

Theoretically it might also be possible to modify the surface of a bacteriophage by incorporating specific protein sequences to preferentially target the particle to particular cell types, e.g. galactose residues to target galactose-recognizing hepatic receptors in the liver [Molenaar et al. 2002] or peptides isolated by screening phage-display libraries to target dendritic [Curiel et al. 2004] or Langerhans cells [McGuire et al. 2004].

To screen phage for the ability to target specific tissue types, phage-display libraries have been passed through mice several times and at each stage phage were isolated from specific tissues [Rajotte et al. 1998]. A similar in vivo screening strategy was also used to isolate phage displaying peptides that showed increased cytoplasmic uptake into mammalian cells [Ivanenkov and Menon 2000]. Phage-displayed peptides so selected may be used as targeted vehicle for antibiotics or anti-tumorals, or act themselves as targeted anti-bacterials and anti-tumorals. Specific phage-displayed peptides could be used, for example, in anticancer therapy either directly inducing apoptosis processes or targeting anti-tumorals to cancer cells, or also targeting to microorganisms that in turn specifically infect tumoral cells.

Although antibodies dominate in most targeting studies, their clinical use may be limited because their large size and high affinity interaction with the first antigen molecules they come in contact with, factors that preclude efficient penetration into tumor, as well as immunogenicity and non-specific uptake by the reticulo-endothelial system. Short peptides demonstrating acceptable affinity and specificity to their targets represent an attractive alternative, possessing impressive tumor penetrating capacity, being not generally recognized by the mononuclear phagocytic system, and being less likely to prime immune responses. In addition, they exhibit a higher activity per mass and greater stability.

Recently, in vivo phage display has been used to analyze the structure and molecular diversity of tumor vasculature and to select tumor-specific antigens which have revealed stage- and type-specific markers of tumor blood vessels [Li et al. 2006; Brown C. K. et al. 2000]. Peptides identified by this approach also work as vehicles to transport cargo therapeutic reagents to tumors. These peptides and their corresponding cellular proteins and ligands may provide molecular tools to selectively target the addresses of tumors and their pathological blood vessels and might increase the efficacy of therapy while decreasing side effects.

Numerous covalent coupling techniques, including the formation of a disulfide bond, crosslinking between primary amines, reactions between a carboxylic acid and primary amine, between maleimide and thiol, between hydrazide andaldehyde, or between a primary amine and free aldehyde can be used for peptide conjugation. These procedures are quite efficient for the preparation of various targeted liposomes on a small scale –for their preliminary laboratory and clinical studies; however, the cost and reproducibility of these derivatives in quality and a quantity sufficient for pharmaceutical applications are challenging problems. These considerations led to evaluate the potential of intact phage fusion coat proteins isolated in biopanning experiments, as easily available targeting components of drug preparations. Using the assembly of the phage proteins with lipid membranes, a model streptavidin-targeted protein was incorporated into 80nm liposomes [Nobs et al. 2006]. The streptavidin-binding phage was affinity selected from the landscape phage library, stripped and incubated with liposomes, which as a result of fusion with the targeted phage proteins acquired a new emergent property –the ability to bind streptavidin, streptavidin-conjugated both fluorescent molecules and gold-beads. When landscape phage serve as a reservoir of the targeted membrane proteins, one of the most troublesome steps of the conjugation technology is bypassed. Furthermore, no reengineering of the selected phage is required at all: the phage themselves serve as the source of the final product –coat protein genetically fused to the targeting peptide. A culture of phage-secreting cells is an efficient, convenient and discontinuous protein production system, and purification of the phage is easily accomplished by simple, routine steps that do not differ from one clone to another.

GENE DELIVERY

The demonstrated ability of targeted fusion filamentous phage to penetrate into mammalian cells, survive inside cellular compartments and express transgenes in target cells has aroused considerable interest in them as potential therapeutic gene delivery vectors [Barry et al. 1996; Dunn 1996]. The phage coat protects the DNA from degradation after injection, and the ability to display foreign molecules on the phage coat also enables targeting of specific cell types, a prerequisite for effective gene therapy. Both artificial covalent conjugation [Larocca et al. 1998] and phage display [Larocca et al. 1999] have been used to display targeting and/or processing molecules on the phage surface. This demonstrates, again, the versatility of phage, showing that tissue targeting can be achieved either by rational design or by the screening of random phage-display libraries.

Unlike animal viruses that have broad natural tropisms, phage have no known natural receptors on mammalian cells [Barrow and Soothill 1997]. Receptor-mediated internalization of phage vectors by mammalian cells can occur if they display foreign cell-specific ligands, such as fibroblast growth factor (FGF2), anti-ErbB2 scFv F5 antibody or integrin-binding peptides [Poul and Marks 1999; Ivanenkov et al. 1999; Larocca et al. 1999; Monaci et al. 2001; Sergeeva et al.

2006]. Phage-derived delivery vehicles are not toxic to mammals and can be manufactured using very simple and safe technologies.

Tumor-specific pVIII- and pIII-fusion phage can be easily transformed into gene delivery system–phagemid infective particles (PIPs) encapsulated within bacteria by phage-encoded peptides [Mount et al. 2004; Li et al. 2005, 2006]. These particles, although significantly smaller than normal phage, have all the elements necessary for delivery and expression of genes in mammalian cells. In this approach, shown first for glioma cells [Mount et al. 2004], a phagemid expressing a model marker or particular therapeutic gene can be easily exchanged for a phagemid expressing a different therapeutic or reporter gene. In addition, a different helper phage selected from a phage display library can target any cell type and direct the encapsulation of any suitable phagemid. Because of its versatility, the PIP system may be readily used for optimization of the gene delivery strategies applied to specific cell and tissue targets.

Phage-derived gene delivery vectors, however, have yet low transformation efficiency. A new approach, demonstrated by Hajitou *et al* [2006], allows dramatic improvements in post-targeting expression of filamentous phage-borne transgenes by genetic hybridization of the phage vector with compatible inverted terminal repeats from adeno-associated virus (AAV). The hybrid vector, termed **AAVP, targeted to α v integrins** by pIII-fused RGD-4C peptide, exhibited much higher mammalian transduction efficiency over corresponding phage-based vectors, probably because of better persistence of episomal DNA in the mammalian cell and the multimerization of the entire transgene cassette during hybrid DNA maintenance. In addition, the AAVP particles have been shown to serve as molecular imaging probes in vivo, allowing monitoring and prediction of drug response in a nude rat model of human sarcoma [Hajitou et al. 2008]. These recent breakthrough advances in the development of targeted phage-derived delivery systems demonstrate their enormous potential for systemic targeted delivery of therapeutic genes and molecular imaging reporter transgenes in solid tumors of cancer patients.

MOLECULAR IMAGING

Modern clinical cancer treatments require a detailed molecular characterization of tumors to optimize the treatment regimen for individual patients. Molecular imaging would allow clinicians to visualize the expression and activity of specific molecules (e.g., proteases and protein kinases) and biological processes (e.g., apoptosis, angiogenesis and metastasis) that influence tumor behaviour and/or response to therapy [Stephen and Gillies 2007; Weissleder 2006]. A general approach to improving the accumulation of the imaging agents at the target site is to conjugate the imaging label to a ligand that binds to a specific molecular target (active probes). The probes bind the targets and are retained at the target site, while unbound probes are cleared from circulation. This approach is most useful for tumor imaging, as cancer cells often overexpress certain surface receptors.

Phage display methods can promote discovery, validation and visualization of molecular markers in a disease-specific way. Recently, disease-specific and organ-specific phage clones were obtained by in vivo phage selection [Krag et al. 2006; Ludke et al. 2007; Pasqualini and Ruoslahti 1996; Rajotte et al. 1998; Valadon et al. 2006], useful as molecular recognition interface for non-invasive in vivo imaging agents [Kelly et al. 2004, 2006a, 2006b; Newton et al. 2006, 2007]. Phage can function as imaging agents as a scaffold for chemical attachment of targeting peptides and imaging labels, as an in vitro or in vivo preselected cancer-specific phage probe -directly conjugated with imaging labels or biotinylated and subsequently imaged with streptavidin conjugated to a radio-active label-, and as a gene-delivery vector with a reporter transgene. Another phage-derived contrast reagent, called 'magnetophage', has been obtained by Segers *et al* [2007] by covalently coupling ultra-small particles of iron oxide and their PEGylated forms to phage capsids. Magnetophages were proposed as a potential magnetic resonance imaging agent, allowing the discrimination of apoptotic cells from normal control cells in a model system. The stealthy PEG-magnetophages, largely invisible to phagocytic cells, were successfully targeted to apoptotic cells of the liver.

The first uses of filamentous phage for imaging of tumors in vivo hold great promise for their future use as bioselective molecular imaging agents in clinical applications. Clinical trials have shown that serial library administration in cancer patients during the screening of phage libraries for cancer-specific ligands can be accomplished without major untoward clinical side effects [Krag et al. 2006].

REFERENCES

Adey, N. B., Mataragnon, A. H., Rider, J. E., Carter, J. M., & Kay, B. K. (1995). Characterization of phage that bind plastic from phage-displayed random peptide libraries. *Gene, 156,* 27–31.

Al-Khaldi, S. F., Mossoba, M. M., Yakes, B. J., Brown, E., Sharma, D., & Carnazza, S. (2008). Recent Advances in microbial discovery using metagenomics, DNA microarray, biosensors, molecular subtyping, and phage recombinant probes. In M. K. Moretti & L. J. Rizzo (Eds.), *Oligonucleotide Array Sequence Analysis* (Chapter 4, pp. 123-147). Nova Science Publishers, Inc.

Al-Khaldi, S. F., Mossoba, M. M., Yakes, B. J., Brown, E., Sharma, D., & Carnazza, S. (2009). The biggest winners in DNA and protein sequence analysis: metagenomics, DNA microarray, biosensors, molecular subtyping, and phage recombinant probes. *International Journal of Medical and Biological Frontiers (Nova Science Publishers), 15(3/4),* 4.

Amstutz, P., Pelletier, J. N., Guggisberg, A., Jermutus, L., Cesaro-Tadic, S., Zahnd, C., & Plückthun, A. (2002). In vitro selection for catalytic activity with ribosome display. *J. Am. Chem. Soc., 124,* 9396-403.

Amstutz, P., Forrer, P., Zahnd, C., & Plückthun, A. (2001). In vitro display technologies: novel developments and applications. *Curr. Opin. Biotechnol., 12,* 400–5.

Arap, W., Kolonin, M. G., Trepel, M., Lahdenranta, J., Cardo-Vila, M., Giordano, R. J., Mintz, P. J., Ardelt, P. U., Yao, V. J., Vidal, C. I., Chen, L., Flamm, A., Valtanen, H., Weavind, L. M., Hicks, M. E., Pollock, R. E., Botz, G. H., Bucana, C. D., Koivunen, E., Cahill, D., Troncoso, P., Baggerly, K. A., Pentz,

R. D., KA D., Logothetis, C. J., & Pasqualini R. (2002). Steps toward mapping the human vasculature by phage display. *Nat. Med., 8,* 121–7.

Arnold, F. H. (1998). Design by directed evolution. *Acc. Chem. Res., 31(3),* 125-31.

Baggio, R., Burgstaller, P., Hale, S. P., Putney, A. R., Lane, M., Lipovsek, D., Wright, M. C., Roberts, R. W., Liu, R., Szostak, J. W., & Wagner, R. W. (2002). Identification of epitope-like consensus motifs using mRNA display. *J. Mol. Recognit., 15,* 126-34.

Balass, M., Heldman, Y., Cabilly, S., Givol, D., Katchalski-Katzir, E., & Fuchs, S. (1993). Identification of a hexapeptide that mimics a conformation-dependent binding site of acetylcholine receptor by use of a phage-epitope library. *Proc. Natl. Acad. Sci. USA, 90,* 10638-42.

Balasubramanian, S., Sorokulova, I. B., Vodyanoy, V. J., & Simonian, A. L. (2007). Lytic phage as a specific and selective probe for detection of *Staphylococcus aureus.* A surface plasmon resonance spectroscopic study. *Biosens. Bioelectr., 22,* 948-55.

Ball, P. (2005). Synthetic biology for nanotechnology. Tutorial. *Nanotechnology, 16,* R1-R8.

Barbas, C. F., Bain, J. D., Hoekstra, D. M., & Lerner, R. A. (1992). Semisynthetic combinatorial libraries: a chemical solution to the diversity problem. *Proc. Natl. Acad. Sci. USA, 89,* 4457–61.

Barbas, C. F., Kang, A. S., Lerner, R. A., & Benkovic, S. J. (1991). Assembly of combinatorial antibody libraries on phage surfaces: the gene III site. *Proc. Natl. Acad. Sci. USA, 88,* 7978-82.

Barrick, J. E., & Roberts, R. W. (2002). Sequence analysis of an artificial family of RNA-binding peptides. *Protein Sci., 11,* 2688-96.

Barrick, J. E., Takahashi, T. T., Balakin, A., & Roberts, R. W. (2001). Selection of RNA-binding peptides using mRNA–peptide fusions. *Methods, 23,* 287-93.

Barrow, P. A., & Soothill, J. S. (1997). Bacteriophage therapy and prophylaxis: rediscovery and renewed assessment of potential. *Trends Microbiol., 5(7),* 268-71.

Barry, M. A., Dower, W. J., & Johnston, S. A. (1996). Toward cell-targeting gene therapy vectors: selection of cell-binding peptides from random peptide-presenting phage libraries. *Nat. Med., 2,* 299–305.

Bass, S., Green, R., & Wells, J. A. (1990). Hormone phage: an enrichment method for variant proteins with altered binding properties. *Proteins: Struct. Funct. Genet., 8,* 309-14.

Belcher, A. M., & Lee, S.-W. (2008). Viral fibers. United States Patent 7332321.

Benhar, I. (2001). Biotechnological applications of phage and cell display. *Biotechnol. Adv., 19,* 1–33.

Berglund, J., Lindbladh, C., Nicholls, I. A., & Mosbach, K. (1998). Selection of phage display combinatorial library peptides with affinity for a yohimbine imprinted methacrylate polymer. *Anal. Commun., 35,* 3–7.

Better, M., Chang, C. P., Robinson, R. R., & Horwitz, A.H. (1988). *Escherichia coli* secretion of an active chimeric antibody fragment. *Science, 240,* 1041-3.

Bi, Q., Cen, X., Wang, W., Zhao, X., Wang, X., Shen, T., & Zhu, S. (2007). A protein microarray prepared with phage-displayed antibody clones. *Biosens. Bioelectron., 22(12),* 3278-82.

Binz, H. K., Amstutz, P., Kohl, A., Stumpp, M. T., Briand, C., Forrer, P., Grütter, M. G., & Plückthun, A. (2004). High-affinity binders selected from designed ankyrin repeat protein libraries. *Nat. Biotechnol., 22,* 575-82.

Binz, H. K., Stumpp, M. T., Forrer, P., Amstutz, P., & Plückthun, A. (2003). Designing repeat proteins: well-expressed, soluble and stable proteins from combinatorial libraries of consensus ankyrin repeat proteins. *J. Mol. Biol., 332,* 489-503.

Bishop-Hurley, S. L., Schmidt, F. J., Erwin, A. L., & Smith, A. L. (2005). Peptides selected for binding to a virulent strain of *Haemophilus influenzae* by phage display are bactericidal. *Antimicrob. Agents Chemother., 49,* 2972-8.

Block, S. M. (1997). Real engines of creation. *Nature, 386,* 217–19.

Boder E. T., Midelfort K. S., & Wittrup K. D. (2000). Directed evolution of antibody fragments with monovalent femtomolar antigen-binding affinity. *Proc. Nat. Acad. Sci. USA, 97(20),* 10701-5.

Boder, E. T., & Wittrup, K. D. (1997). Yeast surface display for screening combinatorial polypeptide libraries. *Nat. Biotech., 15,* 553-57.

Boder, E. T., & Wittrup, K. D. (1998). Optimal screening of surface-displayed polypeptide libraries. *Biotechnol. Prog., 14,* 55-62.

Bratkovic, T., Lunder, M., Popovic, T., Kreft, S., Turk, B., Strukelj, B., & Urleb, U. (2005). Affinity selection to papain yields potent peptide inhibitors of cathepsins L, B, H, and K. *Biochem. Biophys. Res. Commun., 332(3),* 897–903.

Breitling, F., Dübel, S., Seehaus, T., Klewinghaus, I., & Little, M. (1991). A surface expression vector for antibody screening. *Gene, 104,* 147-53.

Brigati, J. R., & Petrenko, V. A. (2005). Thermostability of landscape phage probes. *Anal. Bioanal. Chem., 382,* 1346-50.

Brigati, J., Williams, D. D., Sorokulova, I. B., Nanduri, V., Chen, I. H., Turnbough, C. L., Jr. & Petrenko, V. A. (2004). Diagnostic probes for

Bacillus anthracis spores selected from a landscape phage library. *Clin. Chem., 50,* 1899-906.

Brown, C. K., Modzelewski, R. A., Johnson, C. S., & Wong, M.K. (2000). A novel approach for the identification of unique tumor vasculature bonding peptides using an *E. coli* peptide display library. *Ann. Surg. Oncol. 7(10),* 743-9.

Brown, S. (1992). Engineered iron oxide-adhesion mutants of the *Escherichia coli* phage l receptor. *Proc. Natl Acad. Sci. USA, 89,* 8651-5.

Brown, S. (1997). Metal-recognition by repeating polypeptides. *Nat. Biotechnol., 15,* 269-72.

Brown, S., Sarikaya, M., & Johnson, E. (2000). Genetic analysis of crystal growth. *J. Mol. Biol., 299,* 725-32.

Bryson, J. W., Betz, S. F., Lu, H. S., Suich, D. J., Zhou, H. X., O'Neil, K. T., & DeGrado, W. F. (1995). Protein design: a hierarchic approach. *Science, 270,* 935-41.

Burton, D. R., Barbas, C. F. 3rd, Persson, M. A., Koenig, S., Chanock, R. M., & Lerner, R. A. (1991). A large array of human monoclonal antibodies to type 1 human immunodeficiency virus from combinatorial libraries of asymptomatic individuals. *Proc. Natl. Acad. Sci. USA, 88,* 10134–7.

Cabibbo, A., Sporeno, E., Toniatti, C., Altamura, S., Savino, R., Paonessa, G., & Ciliberto, G. (1995). Monovalent phage display of human interleukin (hIL)-6: Selection of superbinder variants from a complex molecular repertoire in the hIL-6 D-helix. *Gene, 167,* 41-7.

Cai, X., & Garen, A. (1995). Anti-melanoma antibodies from melanoma patients immunized with genetically modified antilogous tumor cells: selection of specific antibodies from single-chain Fv fusion phage libraries. *Proc. Natl. Acad. Sci. USA, 92,* 6537–41.

Carmen, S., & Jermutus, L. (2002). Concepts in antibody phage display. *Brief Funct. Genomic Proteomic, 1(2),* 189-203.

Carnazza, S., Foti, C., Gioffrè, G., Felici, F., & Guglielmino, S. (2008). Specific and selective probes for *Pseudomonas aeruginosa* from phage-displayed random peptide libraries. *Bios. Bioelectron., 23,* 1137-44.

Carnazza, S., Gioffrè, G., Felici, F., & Guglielmino, S. (2007). Recombinant phage probes for *Listeria monocytogenes. J. Phys. Condensed Matter, 19,* 395011.

Cesaro-Tadic, S., Lagos, D., Honegger, A., Rickard, J. H., Partridge, L. J., Blackburn, G. M., & Plückthun, A. (2003). Turnover-based in vitro selection and evolution of biocatalysts from a fully synthetic antibody library. *Nat. Biotechnol., 21,* 679-85.

Chang, C. N., Landolfi, N. F., & Queen, C. (1991). Expression of antibody Fab domains on bacteriophage surfaces. *J. Immunology, 147,* 3610–4.

Chasteen, L., Ayriss, J., Pavlik, P., & Bradbury, A. R. (2006). Eliminating helper phage from phage display. *Nucleic Acids Res., 34(21), e145* [Epub 2006 Nov 6].

Chen, B. X., Wilson, S. R., Das, M., Coughlin, D. J., & Erlanger, B. F. (1998). Antigenicity of fullerenes: antibodies specific for fullerenes and their characteristics. *Proc. Natl. Acad. Sci. USA, 95,* 10809-13.

Chen, Y. (2003). *Novel approach to generate human monoclonal antibodies by PROfusion^{TM} Technology*. Cambridge Healthtech Institute 4th Annual Conference on Recombinant Antibodies. Cambridge, MA, USA.

Cheng, X., Kay, B. K., & Juliano, R. L. (1996). Identification of a biologically significant DNA-binding peptide motif by use of a random phage display library. *Gene, 171,* 1–8.

Chiang, Y. L., Sheng-Dong, R., Brow, M. A., & Larrick, J. W. (1989). Direct cDNA cloning of the rearranged immunoglobulin variable region. *Biotechniques, 7,* 360–6.

Chiarabelli, C., Vrijbloed, J. W., Thomas, R. M., & Luisi, P. L. (2006a). Investigation of de novo totally random biosequences. Part I. A general method for in vitro selection of folded domains from a random polypeptide library displayed on phage. *Chem. Biodiv., 3,* 827-39.

Chiarabelli, C., Vrijbloed, J. W., De Lucrezia, D., Thomas, R. M., Stano, P., Polticelli, F., Ottone, T., Papa, E., & Luisi, P. L. (2006b). Investigation of de novo totally random biosequences. Part II. On the folding frequency in a totally random library of de novo proteins obtained by phage display. *Chem. Biodiv., 3,* 840-59.

Chopra, P., & Kamma, A. (2006). Engineering life through Synthetic Biology. *In Silico Biology, 6,* 401-10.

Chowdhury, P. S. (2002). Targeting random mutations to hotspots in antibody variable domains for affinity improvement. In P. M. O'Brien, & R. Aitken (Eds.), *Antibody Phage Display: Methods and Protocols.* (pp. 269–86). Totowa, NJ: Humana Press.

Clackson, T., Hoogenboom, H. R., Griffiths, A. D., & Winter, G. (1991). Making antibody fragments using phage display libraries. *Nature, 352,* 624–8.

Clark, J. R., & March, J. B. (2004a). Bacterial viruses as human vaccines? *Expert Rev. Vaccines, 3,* 463–76.

Clark, J. R., & March, J. B. (2004b). Bacteriophage-mediated nucleic acid immunization. *FEMS Immunol. Med. Microbiol., 40,* 21–6.

Clark, J. R., & March, J. B. (2006). Bacteriophages and biotechnology: vaccines, gene therapy and antibacterials. *Trends Biotech., 24,* 212-8.

Coley, A. M., Campanale, N. V., Casey, J. L., Hodder, A. N., Crewther, P. E., Anders, R. F., Tilley, L. M., & Foley, M. (2001). Rapid and precise epitope mapping of monoclonal antibodies against *Plasmodium falciparum* AMA1 by combined phage display of fragments and random peptides. *Protein Eng., 14,* 691–8.

Cortese, R., Monaci, P., Nicosia, A., Luzzago, A., Felici, F., Galfre, G., Pessi, A., Tramontano, A., & Sollazzo, M. (1995). Identification of biologically active peptides using random libraries displayed on phage. *Curr. Opin. Biotechnol., 6,* 73-80.

Cui, X., Hetke, J. F., Wiler, J. A., Anderson, D. J., & Martin, D. C. (2001). Electrochemical deposition and characterization of conducting polymer polypyrrole/PSS on multichannel neural probes. *Sensors Actuat. A, 93,* 8–18.

Cujec, T. P., Medeiros, P. F., Hammond, P., Rise, C., & Kreider, B. L. (2002). Selection of v-abl tyrosine kinase substrate sequences from randomized peptide and cellular proteomic libraries using mRNA display. *Chem. Biol., 9,* 253-264.

Curiel, T. J., Morris, C., Brumlik, M., Landry, S. J., Finstad, K., Nelson, A., Joshi, V., Hawkins, C., Alarez, X., Lackner, A., & Mohamadzadeh, M. (2004). Peptides identified through phage display direct immunogenic antigen to dendritic cells. *J. Immunol., 172,* 7425–31.

Cwirla, S. E., Peters, E. A., Barrett, R. W., & Dower, W. J. (1990). Peptides on phage: a vast library of peptides for identifying ligands. *Proc. Natl. Acad. Sci. USA, 87,* 6378-82.

Dall'Acqua, W., & Carter, P. (1998). Antibody engineering. *Curr. Opin. Struct. Biol., 8,* 443-50.

De Lucrezia, D., Franchi, M., Chiarabelli, C., Gallori, E., & Luisi, P. L. (2006a). Investigation of de novo totally random biosequences. Part III. RNA foster: a novel assay to investigate RNA folding structural properties. *Chem. Biodiv., 3,* 860-8.

De Lucrezia, D., Franchi, M., Chiarabelli, C., Gallori, E., & Luisi, P. L. (2006b). Investigation of de novo totally random biosequences. Part IV. Folding properties of de novo, totally random RNAs. *Chem. Biodiv., 3,* 869-77.

Deming, T. J. (1997). Polypeptide materials: new synthetic methods and applications. *Adv. Mater., 9,* 299-311.

Deng, S. J., MacKenzie, C. R., Sadowska, J., Michniewicz, J., Young, N. M., Bundle, D. R., & Narang, S. A. (1994). Selection of antibody single-chain variable fragments with improved carbohydrate binding by phage display. *J. Biol. Chem., 269*, 9533–8.

Dennis, M. S., Eigenbrot, C., Skelton, N. J., Ultsch, M. H., Santell, L., Dwyer, M. A., O'Connell, M. P., & Lazarus, R. A. (2000). Peptide exosite inhibitors of factor VIIa as anticoagulants. *Nature, 404*, 465-70.

Dente, L., Cesareni, G., Micheli, G., Felici, F., Folgori, A., Luzzago, A., Monaci, P., Nicosia, A. & Delmastro, P. (1994). Monoclonal antibodies that recognize filamentous phage: tools for phage display technology. *Gene, 148(1)*, 7-13.

Dickerson, T. J., Kaufmann, G. F., & Janda, K. D. (2005). Bacteriophage-mediated protein delivery into the central nervous system and its application in immunopharmacotherapy. *Expert Opin. Biol. Ther., 5*, 773–81.

Dultsev, F. N., Speight, R. E., Fiorini, M. T., Blackburn, J. M., Abell, C., Ostanin, V. P., & Klenerman, D. (2001). Direct and quantitative detection of bacteriophage by "hearing" surface detachment using a quartz crystal microbalance. *Anal. Chem., 73(16)*, 3935-9.

Dunn, I. S. (1996). Mammalian cell binding and transfection mediated by surface-modified bacteriophage lambda. *Biochimie, 78*, 856–61.

Dwyer, M. A., Looger, L. L., & Hellinga, H. W. (2004). Computational design of a biologically active enzyme. *Science, 304*, 1967-71.

El-Mousawi, M., Tchistiakova, L., Yurchenko, L., Pietrzynski, G., Moreno, M., Stanimirovic, D., Ahmad, D., & Alakhov, V. (2003). A vascular endothelial growth factor high affinity receptor 1-specific peptide with antiangiogenic activity identified using a phage display peptide library. *J. Biol. Chem., 278*, 46681–91.

Emanuel, P., O'Brien, T., Burans, J., DasGupta, B. R., Valdes, J. J., & Eldefrawi, M. (1996). Directing antigen specificity towards botulinum neurotoxin with combinatorial phage display libraries. *J. Immunol. Methods, 193*, 189–97.

Erlanger, B. F., Chen, B.-X., Zhu, M., & Brus, L. (2001). Binding of an anti-fullerene IgG monoclonal antibody to single wall carbon nanotubes. *Nano Lett., 1*, 465-7.

Fakok, V. A., Bratton, D. L., Rose, D. M., Pearson, A., Ezekewitz, R. A. B., & Henson, P. M. (2000). A receptor for phosphatidylserine-specific clearance of apoptotic cells. *Nature, 405*, 85-90.

Feldhaus, M. J., & Siegel, R. W. (2004a). Flow cytometric screening of yeast surface display libraries. *Methods Mol. Biol., 263*, 311-32.

Feldhaus, M. J., & Siegel, R. W. (2004b). Yeast display of antibody fragments: A discovery and characterization platform. *J. Immunol. Methods, 290*, 69-80.

Felici, F., Luzzago, A., Monaci, P., Nicosia, A., Sollazzo, M., & Traboni, C. (1995). Peptide and protein display on the surface of filamentous bacteriophage. In M. R. El-Gewely (Ed.), *Biotechnology Annual Review* (Volume 1, 149-83). Amsterdam, The Netherlands: Elsevier Science B.V.

Finucane, M. D., Tuna, M., Lees, J. H., & Woolfson, D. N. (1999). Core-directed protein design. I. An experimental method for selecting stable proteins from combinatorial libraries. *Biochemistry, 38,* 11604-12.

Flynn, C. E., Mao, C., Hayhurst, A., Williams, J. L., Georgiou, G., Iversona B., & Belcher, A. M. (2003). Synthesis and organization of nanoscale semiconductor materials using evolved peptide specificity and viral capsid assembly. *J. Mater. Chem., 13,* 2414–21.

Folgori, A., Tafi, R., Meola, A., Felici, F., Galfré, G., Cortese, R., Monaci, P., & Nicosia, A. (1994). A general strategy to identify mimotopes of pathological antigens using only random peptide libraries and human sera. *EMBO J., 13,* 2236–43.

Francisco, J. A., Campbell, R., Iverson, B. L., & Georgiou, G. (1993). Production and fluorescence-activated cell sorting of *Escherichia coli* expressing a functional antibody fragment on the external surface. *Proc. Nat. Acad. Sci. USA, 90,* 10444-8.

Fu, L., Li, S., Zhang, K., Chen, I.-H., Petrenko, V. A., & Cheng, Z. (2007). Magnetostrictive microcantilever as an advanced transducer for biosensors. *Sensors J., 7,* 2929-41.

Garrard, L. J., Yang, M., O'Connell, M. P., Kelley, R. F., & Henner, D. J. (1991). Fab assembly and enrichment in a monovalent phage display system. *Bio/Technology, 9,* 1373-7.

Georgiou, G., Stathopoulos, C., Daugherty, P. S., Nayak, A. R., Iverson, B. L., & Curtis, R. III. (1997). Display of heterologous proteins on the surface of microorganisms: from the screening of combinatorial libraries to live recombinant vaccines. *Nat. Biotech., 15,* 29-34.

Geysen, H. M., Rodda, S. J., & Mason, T. J. (1986a). A priori delineation of a peptide which mimics a discontinuous antigenic determinant. *Mol. Immunol., 23,* 709-15.

Geysen, H. M., Rodda, S. J., & Mason, T. J. (1986b). The delineation of peptides able to mimic assembled epitopes. *Ciba Found. Symp., 119,* 130-49.

Giordano, R. J., Cardo-Vila, M., Lahdenranta, J., Pasqualini, R., & Arap, W. (2001). Biopanning and rapid analysis of selective interactive ligands. *Nat. Med., 7,* 1249–53.

Gold, L. (2001). mRNA display: diversity matters during in vitro selection. *Proc. Natl. Acad. Sci. USA, 98(9),* 4825-6.

Goldman, E. R., Pazirandeh, M. P., Mauro, J. M., King, K. D., Frey, J. C., & Anderson, G. P. (2000). Phage-displayed peptides as biosensor reagents. *J. Mol. Recognit., 13,* 382-7.

Gough, K.C., Cockburn, W., & Whitelam, G.C. (1999). Selection of phage-display peptides that bind to cucumber mosaic virus coat protein. *J. Virol. Methods, 79,* 169–80.

Graff, C. P., Chester, K., Begent, R., & Wittrup, K. D. (2004). Protein Engineering Design and Selection. *Prot. Eng. Des. Sel., 17,* 293-304.

Graus, Y. F., de Baets, M. H., Parren, P. W., Berrih-Aknin, S., Wokke, J., van Breda Vriesman, P. J., & Burton, D. R. (1997). Human anti-nicotinic acetylcholine receptor recombinant Fab fragments isolated from thymus-derived phage display libraries from myasthenia gravis patients reflect predominant specificities in serum and block the action of pathogenic serum antibodies. *J. Immunol., 158,* 1919–29.

Greenwood, J., Willis, A.E., & Perham, R.N. (1991). Multiple display of foreign peptides on a filamentous bacteriophage. *J. Mol. Biol., 220,* 821-7.

Griffiths, A. D., & Duncan, A. R. (1998). Strategies for selection of antibodies by phage display. *Curr. Opin. Biotechnol., 9,* 102-8.

Gu, H. D., Yi, Q. A., Bray, S. T., Riddle, D. S., Shiau, A. K., & Baker, D. (1995). A phage display system for studying the sequence determinants of protein-folding. *Protein Science, 4,* 1108-17.

Hajitou, A., Lev, D. C., Hannay, J. A., Korchin, B., Staquicini, F. I., Soghomonyan, S., Alauddin, M. M., Benjamin, R. S., Pollock, R. E., Gelovani, J. G., Pasqualini, R., & Arap, W. (2008). A preclinical model for predicting drug response in soft-tissue sarcoma with targeted AAVP molecular imaging. *Proc. Natl. Acad. Sci. USA, 105(11),* 4471-6.

Hajitou, A., Trepel, M., Lilley, C. E., Soghomonyan, S., Alauddin, M. M., Marini, F. C. 3rd, Restel, B. H., Ozawa, M. G., Moya, C. A., Rangel, R., Sun, Y., Zaoui, K., Schmidt, M., von Kalle, C., Weitzman, M. D., Gelovani, J. G., Pasqualini, R., & Arap, W. (2006). A hybrid vector for ligand-directed tumor targeting and molecular imaging. *Cell, 125(2),* 385-98.

Hammond, P. W., Alpin, J., Rise, C. E., Wright, M., & Kreider, B. L. (2001). In vitro selection and characterization of Bcl-X(L)-binding proteins from a mix of tissue-specific mRNA display libraries. *J. Biol. Chem., 276,* 20898-906.

Hanes, J., & Plückthun, A. (1997). In vitro selection and evolution of functional proteins using ribosome display. *Proc. Natl. Acad. Sci. USA, 94,* 4937-42.

Hanes, J., Jermutus, L., & Plückthun, A. (2000a). Selecting and evolving functional proteins in vitro by ribosome display. *Methods Enzymol., 328,* 404-30.

Hanes, J., Schaffitzel, C., Knappik, A., & Plückthun, A. (2000b). Picomolar affinity antibodies from a fully synthetic naïve library selected and evolved by ribosome display. *Nat. Biotechnol., 18,* 1287-92.

He, M., & Taussig, M. J. (1997). Antibody-ribosome-mRNA (ARM) complexes as efficient selection particles for in vitro display and evolution of antibody combining sites. *Nucleic Acids Res., 25,* 5132-4.

He, M., & Taussig, M. J. (2007). Eukaryotic Ribosome Display with in situ DNA recovery. *Nat. Methods, 4,* 281-8.

Hessling, J., Lohse, M. J., & Klotz, K. N. (2003). Peptide G protein agonists from a phage display library. *Biochem. Pharmacol., 65,* 961–7.

Hong, L., Liao, W., Zhao, X. S., & Zhu, S. G. (2004). Phage antibody chip for discriminating proteomes from different cells. *Acta Phys.-Chim. Sin., 20,* 1182-5.

Hoogenboom, H. R. (1997). Designing and optimizing library selection strategies for generating high-affinity antibodies. *Trends Biotechnol., 15,* 62–70.

Hoogenboom, H. R., & Chames, P. (2000). Natural and designer binding sites made by phage display technology. *Immunol. Today, 21,* 371-8.

Hoogenboom, H. R., de Bruïne, A. P., Hufton, S. E., Hoet, R. M., Arends, J. W., & Roovers, R. C. (1998). Antibody phage display technology and its applications. *Immunotechnology, 4,* 1–20.

Hoogenboom, H. R., Griffiths, A. D., Johnson, K. S., Chiswell, D. J., Hudson, P., & Winter, G. (1991). Multi-subunit proteins on the surface of filamentous phage: methodologies for displaying antibody (Fab) heavy and light chains. *Nucl. Acids Res., 19,* 4133-7.

Hooker, J. M., Kovacs, E. W., & Francis, M. B. (2004). Interior surface modification of bacteriophage MS2. *J. Am. Chem. Soc., 126,* 3718-9.

Houimel, M., Schneider, P., Terskikh, A., Mach, J. P. Starovasnik, M. A. & Lowman, H. B. (2002). Stable "zeta" peptides that act as potent antagonists of the high-affinity IgE receptor. *Proc. Natl. Acad. Sci. USA, 99,* 1303–8.

Huang, Y., Chiang, C.-Y., Lee, S. K., Gao, Y., Hu, E. L., De Yoreo, J., & Belcher, A. M. (2005). Programmable assembly of nanoarchitectures using genetically engineered viruses. *Nano Letters, 5(7),* 1429-34.

Huang, L., Sexton, D. J., Skogerson, K., Devlin, M., Smith, R., Sanyal, I., Parry, T., Kent, R., Enright, J., Wu, Q. L., Conley, G., DeOliveira, D., Morganelli, L., Ducar, M., Wescott, C. R., & Ladner, R. C. (2003). Novel peptide inhibitors of angiotensin-converting enzyme 2. *J. Biol. Chem., 278,* 15532–40.

Hust, M., & Dubel, S. (2004). Mating antibody phage display with proteomics. *Trends in Biotechnology, 22,* 8-14

Hyde-DeRuyscher, R., Paige, L. A., Christensen, D. J., Hyde-DeRuyscher, N., Lim, A., Fredericks, Z. L., Kranz, J., Gallant, P., Zhang, J., Rocklage, S. M., Fowlkes, D. M., Wendler, P. A., & Hamilton, P. T. (2000). Detection of small-molecule enzyme inhibitors with peptides isolated from phage-displayed combinatorial peptide libraries. *Chem Biol, 7,* 17-25.

Iqbal, S. S., Mayo, M. W., Bruno, J. G., Bronk, B. V., Batt, C. A., & Chambers, J. P. (2000). A review of molecular recognition technologies for detection of biological threat agents. *Biosens. Bioelectron., 15,* 549–78.

Irving, R. A., Coia, G., Roberts, A., Nuttall, S. D., & Hudson, P. J. (2001). Ribosome display and affinity maturation: from antibodies to single V-domains and steps towards cancer therapeutics. *J. Immunol. Methods, 248,* 31-45.

Isalan, M., & Choo, Y. (2000). Engineered zinc finger proteins that respond to DNA modification by HaeII and HhaI methyltransferase enzymes. *J. Mol. Biol., 295,* 471-7.

Ivanenkov, V. V., Felici, F., & Menon, A. G. (1999). Targeted delivery of multivalent phage display vectors into mammalian cells. *Biochim. Biophys. Acta,* 1448(3), 463-72.

Ivanenkov, V. V., & Menon, A. G. (2000). Peptide-mediated transcytosis of phage display vectors in MDCK cells. *Biochem. Biophys. Res. Commun., 276,* 251–7.

Izhaky, D., & Addadi, L. (1998). Pattern recognition of antibodies for two-dimensional arrays of molecules. *Adv. Mater., 10,* 1009-13.

Jepson, C. D., & March, J. B. (2004). Bacteriophage lambda is a highly stable DNA vaccine delivery vehicle. *Vaccine, 22,* 2413–9.

Jermutus, L., Honegger, A., Schwesinger, F., Hanes, J., & Plückthun, A. (2001). Tailoring in vitro evolution for protein affinity or stability. *Proc. Natl. Acad. Sci. USA, 98,* 75-80.

Johnson, M. L., Wan, J., Huang, S., Cheng, Z., Petrenko, V. A., Kim, D. J., Chen, I. H., Barbaree, J. M., Hong, J. W., & Chin, B. A. (2008). A wireless biosensor using microfabricated phage-interfaced magnetoelastic particles. *Sensors Actuators A. Phys., 144(1),* 38-47.

Kassner, P. D., Burg, M. A., Baird, A., & Larocca D. (1999). Genetic selection of phage engineered for receptor-mediated gene transfer to mammalian cells. Biochem. Biophys. *Res. Commun., 264(3),* 921-8.

Kehoe, J. W., & Kay, B. K. (2005). Filamentous phage display in the new millennium. *Chem. Rev., 105(11),* 4056–72.

Kelly, K., Alencar, H., Funovics, M., Mahmood, U., & Weissleder, R. (2004). Detection of invasive colon cancer using a novel, targeted, library-derived fluorescent peptide. *Cancer Res., 64 (17)*, 6247-51.

Kelly, K. A., Clemons, P. A., Yu, A. M., & Weissleder, R. (2006). High-throughput identification of phage-derived imaging agents. *Mol. Imaging, 5(1)*, 24-30.

Kelly, K. A., Waterman, P., & Weissleder, R. (2006). In vivo imaging of molecularly targeted phage. *Neoplasia, 8(12)*, 1011-8.

Khalil, A. S., Ferrer, J. M., Brau, R. R., Kottmann, S. T., Noren, C. J., Lang, M. J., & Belcher, A. M. (2007). Single M13 bacteriophage tethering and stretching. *PNAS, 104(12)*, 4892-7.

Knappik, A., Ge, L., Honegger, A., Pack, P., Fischer, M., Wellnhofer, G., Hoess, A., Wolle, J., Plückthun, A., & Virnekas, B. (2000). Fully synthetic human combinatorial antibody libraries (HuCAL) based on modular consensus frameworks and CDRs randomized with trinucleotides. *J. Mol. Biol., 296*, 57-86.

Knurr, J., Benedek, O., Heslop, J., Vinson, R. B., Boydston, J. A., McAndrew, J., Kearney, J. F., & Turnbough, C. L. (2003). Peptide ligands that bind selectively to spores of *Bacillus subtilis* and closely related species. *Appl. Environ. Microbiol., 69*, 6841–7.

Koide, A., Bailey, C. W., Huang, X., & Koide, S. (1998). The fibronectin type III domain as a scaffold for novel binding proteins. *J. Mol. Biol., 284*, 1141-51.

Konturri, K., Pentti, P., & Sundholm, G. (1998). Polypyrrole as a model membrane for drug delivery. *J. Electroanal. Chem., 453*, 231–8.

Kouzmitcheva, G. A., Petrenko, V. A., & Smith, G. P. (2001). Identifying diagnostic peptides for lyme disease through epitope discovery. *Clin. Diagn. Lab. Immunol., 8*, 150-60.

Krag, D. N., Shukla, G. S., Shen, G.-P., Pero, S., Ashikaga, T., Fuller, S., Weaver, D. L., Burdette-Radoux, S., & Thomas, C. (2006). Selection of tumor-binding ligands in cancer patients with phage display libraries. *Cancer Res., 66(15)*, 7724-33.

Kretzschmar, T. & von Ruden, T. (2002). Antibody discovery: phage display. *Curr. Opin. Biotechnol., 13*, 598-602.

Kristensen, P., & Winter, G. (1998). Proteolytic selection for protein folding using filamentous bacteriophages. *Fold. Des., 3*, 321–8.

Ladner, R. C. (1995). Constrained peptides as binding entities. *Trends Biotechnol., 13*, 426–30.

Ladner, R. C., Sato, A. K., Gorzelany, J., & de Souza, M. (2004). Phage display-derived peptides as therapeutic alternatives to antibodies. *Drug Discov. Today, 9,* 525–9.

Lakshmanan, R. S., Guntupalli, R., Hu, J., Kim, D.-J., Petrenko, V. A., Barbaree, J. M., & Chin, B. A. (2007). Phage immobilized magnetoelastic sensor for the detection of *Salmonella typhimurium. J. Microbiol. Methods, 71(1),* 55-60.

Langer, R., & Tirrell, D. A. (2004). Designing materials for biology and medicine. *Nature, 428,* 487–92.

Larocca, D., Kassner, P. D., Witte, A., Ladnera, R. C., Pierce, G. F., & Baird, A. (1999). Gene transfer to mammalian cells using genetically targeted filamentous bacteriophage. *FASEB J., 13,* 727–34.

Larocca, D., Witte, A., Johnson, W., Pierce, G. F., & Baird, A. (1998). Targeting bacteriophage to mammalian cell surface receptors for gene delivery. *Hum. Gene Ther., 9,* 2393–9.

LaVan, D. A., Lynn, D. M., & Langer, R. (2002). Moving smaller in drug discovery and delivery. *Nat. Rev. Drug. Discov., 1,* 77–84.

Lee, L.., Buckley, C., Blades, M. C., Panayi, G., George, A. J., & Pitzalis, C. (2002). Identification of synovium-specific homing peptides by in vivo phage display selection. *Arthritis Rheum., 48(8),* 2109-20.

Lee, S.-W., Mao, C., Flynn, C. E., & Belcher, A. M. (2002). Ordering of quantum dots using genetically engineered viruses. *Science, 296,* 892-5.

Legendre, D., & Fastrez, J. (2002). Construction and exploitation in model experiments of functional selection of a landscape library expressed from a phagemid. *Gene, 290,* 203–15.

Leinonen, J., Wu, P., & Stenman, U. H. (2002). Epitope mapping of antibodies against prostate-specific antigen with use of peptide libraries. *Clin. Chem., 48,* 2208–16.

Li, X. B., Schluesener, H. J., & Xu, S. Q. (2006). Molecular addresses of tumors: selection by in vivo phage display. *Arch. Immunol. Ther. Exp. (Warsz), 54(3),* 177-81.

Li, X. B., Schluesener, H. J., & Xu, S. Q. (2006). Molecular addresses of tumors: selection by in vivo phage display. *Arch. Immunol. Ther. Exp. (Warsz), 54(3),* 177-81.

Li, Z., Jiang, H., Zhan, J., & Gu J. (2006). Cell-targeted phagemid particles preparation using Escherichia coli bearing ligand-pIII encoding helper phage genome. *Biotechniques, 41(6),* 706-7.

Li, Z., Zhang, J., Zhao, R., Xu, Y., & Gu, J. (2005). Preparation of peptide-targeted phagemid particles using a protein III-modified helper phage. *Biotechniques, 39(4),* 493-97.

Liao, W., Hong, L., Wei, F., Zhu, S. G., & Zhao, X. S. (2005). Improving phage antibody chip by pVIII display system. *Acta Physico-Chimica Sinica, 21,* 508-11.

Lipovsek, D., & Plückthun, A. (2004). In-vitro protein evolution by ribosome display and mRNA display. *J. Imm. Methods, 290,* 51-67.

Liu, B., & Marks, J. D. (2000). Applying phage antibodies to proteomics: selecting single chain Fv antibodies to antigens blotted on nitrocellulose. *Anal. Biochem., 286,* 119–28.

Liu, F., Luo, Z., Ding, X., Zhu, S. & Yu, X. (2008). Phage-displayed protein chip based on SPR sensing. *Sens. Actuators B: Chem., doi:10.1016/j.snb.2008.11.031.*

Livnah, O., Stura, E. A., Johnson, D. L., Middleton, S. A., Mulcahy, L. S., Wrighton, N. C., Dower, W. J., Jolliffe, L. K., & Wilson, I. A. (1996). Functional mimicry of a protein hormone by a peptide agonist: the EPO receptor complex at 2.8 Å. *Science, 273,* 464–71.

Ludtke, J. J., Sololoff, A. V., Wong, S. C., Zhang, G., & Wolff, J. A. (2007). In vivo selection and validation of liver-specific ligands using a new T7 phage peptide display system. *Drug Deliv., 14(6),* 357-69.

Luisi, P. L. (2007). Chemical aspects of synthetic biology. *Chemistry and Biodiversity, 4,* 603-21.

Lunder, M., Bratkovic, T., Doljak, B., Kreft, S., Urleb, U., Strukelj, B., & Plazar, N. (2005a). Comparison of bacterial and phage display peptide libraries in search of target-binding motif. *Appl. Biochem. Biotechnol., 127(2),* 125–31.

Lunder, M., Bratkovic, T., Kreft, S., & Strukelj, B. (2005b). Peptide inhibitor of pancreatic lipase selected by phage display using different elution strategies. *J. Lipid Res., 46(7),* 1512–6.

Luzzago, A., & Felici, F. (1998). Construction of disulfide-constrained random peptide libraries displayed on phage coat protein VIII. *Methods Mol. Biol., 87,* 155-64.

Mao, C., Flynn, C. E., Hayhurst, A., Sweeney, R., Qi, J., Georgiou, G., Iverson, B., & Belcher, A. M. (2003). Viral assembly of oriented quantum dot nanowires. *Proc. Natl Acad. Sci. USA, 100,* 6946-51.

Mao, C., Solis, D. J., Reiss, B. D., Kottmann, S. T., Sweeney, R. Y., Hayhurst, A., Georgiou, G., Iverson, B., & Belcher, A. M. (2004). Virus-Based toolkit for the directed synthesis of magnetic and semiconducting nanowires. *Science, 303,* 213-7.

March, J. B., Clark, J. R., & Jepson, C. D. (2004). Genetic immunization against hepatitis B using whole bacteriophage lambda particles. *Vaccine, 22,* 1666–71.

Marks, J. D., Hoogenboom, H. R., Bonnert, T. P., McCafferty, J., Griffiths, A. D., & Winter, G. (1991). By-passing immunization: human antibodies from V-gene libraries displayed on phage. *J. Mol. Biol., 222,* 581–97.

Marvin, D.A. (1998). Filamentous phage structure, infection and assembly. *Curr. Opin. Struct. Biol., 8,* 150-8.

Mattheakis, L. C., Bhatt, R. R., & Dower, W.J. (1994). An in vitro polysome display system for identifying ligands from very large peptide libraries. *Proc. Natl. Acad. Sci. USA, 91,* 9022-6.

Maun, H. R., Eigenbrot, C., & Lazarus, R. A. (2003). Engineering exosite peptides for complete inhibition of factor VIIa using a protease switch with substrate phage. *J. Biol. Chem., 278,* 21823–30.

McCafferty, J., Griffiths, A. D., Winter, G., & Chiswell, D. J. (1990). Phage antibodies: filamentous phage displaying antibody variable domains. *Nature, 348,* 552-4.

McConnell, S. J., Dinh, T., Le, M. H., Brown, S. J., Becherer, K., Blumeyer, K., Kautzer, C., Axelrod, F., & Spinella, D. G. (1998). Isolation of erythropoietin receptor agonist peptides using evolved phage libraries. *Biol. Chem., 379,* 1279–86.

McGrath, K. P., Fournier, M. J., Mason, T. L., & Tirrell, D. A. (1992). Genetically directed synthesis of new polymeric materials. Expression of artificial genes encoding proteins with repeating -(AlaGly)3ProGluGly-elements. *J. Am. Chem. Soc., 114,* 727-33.

McGuire, M. J., Sykes, K. F., Samli, K. N., Timares, L., Barry, M. A., Stemke-Hale, K., Tagliaferri, F., Logan, M., Jansa, K., Takashima, A., Brown, K. C., & Johnston, S. A. (2004). A library-selected, Langerhans cell targeting peptide enhances an immune response. *DNA Cell Biol., 23,* 742–52.

McLafferty, M. A., Kent, R. B., Ladner, R. C., & Markland, W. (1993). M13 bacteriophage displaying disulfide-constrained microproteins. *Gene, 128,* 29–36.

Meola, A., Delmastro, P., Monaci, P., Luzzago, A., Nicosia, A., Felici, F., Cortese, R., & Galfre, G. (1995). Derivation of vaccines from mimotopes. Immunologic properties of human hepatitis B virus surface antigen mimotopes displayed on filamentous phage. *J. Immunol., 154,* 3162–72.

Minenkova, O. O., Ilyichev, A. A., Kishchenko, G. P., & Petrenko, V. A. (1993). Design of specific immunogens using filamentous phage as the carrier. *Gene, 128,* 85-88.

Molenaar, T. J., Michon, I., de Haas, S. A. M., van Berkel, T. J. C., Kuiper, J., & Biessen, E. A. L. (2002). Uptake and processing of modified bacteriophage M13 in mice: implications for phage display. *Virology, 293,* 182–91.

Monaci, P., Urbanelli, L., & Fontana, L. (2001). Phage as gene delivery vectors. *Curr. Opin. Mol. Ther., 3(2),* 159-69.

Mount, J. D., Samoylova, T. I., Morrison, N.E., Cox, N. R., Baker, H. J., & Petrenko, V. A. (2004). Cell targeted phagemid rescued by preselected landscape phage. *Gene, 341,* 59-65.

Myers, M. A., Davies, J. M., Tong, J. C., Whisstock, J., Scealy, M., Mackay, I. R., & Rowley, M. J. (2000). Conformational epitopes on the diabetes autoantigen GAD65 identified by peptide phage display and molecular modelling. *J. Immunol., 165,* 3830–8.

Nakamura, G. R., Reynolds, M. E., Chen, Y. M., Starovasnik, M. A., & Lowman, H. B. (2002). Stable "zeta" peptides that act as potent antagonists of the high-affinity IgE receptor. *Proc. Natl. Acad. Sci. USA, 99,* 1303–8.

Nakamura, G. R., Starovasnik, M. A., Reynolds, M. E., & Lowman, H. B. (2001). A novel family of hairpin peptides that inhibit IgE activity by binding to the high-affinity IgE receptor. *Biochemistry, 40,* 9828–35.

Nakashima, T., Ishiguro, N., Yamaguchi, M., Yamauchi, A., Shima, Y., Nozaki, C., Urabe, I., & Yomo, T. (2000). Construction and characterization of phage libraries displaying artificial proteins with random sequences. *J. Biosci. Bioeng., 90,* 253–9.

Nam, K. T., Peelle, B. R., Lee, S.-W., & Belcher, A. M. (2004). Genetically driven assembly of nanorings based on the M13 virus. *Nano Lett., 4,* 23-7.

Nanduri, V., Balasubramanian, S., Sista, S., Vodyanoy, V. J., & Simonian, A. L. (2007a). Highly sensitive phage-based biosensor for the detection of β - galactosidase. *Anal. Chim. Acta, 589,* 166-72.

Nanduri, V., Bhunia, A. K., Tu, S.-I., Paoli, G. C., & Brewster, J. D. (2007b). SPR biosensor for the detection of *L. monocytogenes* using phage-displayed antibody. *Biosens. Bioelectr., 23,* 248-52.

Nanduri, V., Sorokulova, I. B., Samoylov, A. M., Simonian, A. L., Petrenko, V. A., & Vodyanoy, V. (2007c). Phage as a molecular recognition element in biosensors immobilized by physical adsorption. *Biosens. Bioelectron., 22,* 986-92.

Nemoto, N., Miyamoto-Sato, E., Husimi, Y., & Yanagawa, H. (1997). In vitro virus: bonding of mRNA bearing puromycin at the 3'-terminal end to the C-terminal end of its encoded protein on the ribosome in vitro. *FEBS Lett., 414,* 405-8.

Newton, J. R., Kelly, K. A., Mahmood, U., Weissleder, R., & Deutscher, S. L. (2006). In vivo selection of phage for the optical imaging of PC-3 human prostate carcinoma in mice. *Neoplasia, 8(9),* 772-80.

Newton, J. R., Miao, Y., Deutscher, S. L., & Quinn, T. P. (2007). Melanoma imaging with pretargeted bivalent bacteriophage. *J. Nucl. Med. 48 (3),* 429-36.
Nobs, L., Buchegger, F., Gurny, R., & Allemann, E. (2006). Coupling methods to obtain ligand-targeted liposomes and nanoparticles. *Drugs Pharm. Sci., 158,* 123-148.
Ohlin, M., Owman, H., Mach, M., & Borrebaeck, C. A. (1996). Light chain shuffling of a high affinity antibody results in a drift in epitope recognition. *Mol. Immunol., 33,* 47–56.
Olsen, E. V., Sorokulova, I. B., Petrenko, V. A., Chen, I. H., Barbaree, J. M., & Vodyanoy, V. J. (2006). Affinity-selected filamentous bacteriophage as a probe for acoustic wave biodetectors of *Salmonella typhimurium. Biosens. Bioelectron., 21(8),* 1434-42.
Olsen, E. V., Sykora, J. C., Sorokulova, I. B., Chen, I. H., Neely, W. C., Barbaree, J. M., Petrenko, V. A., & Vodyanoy, V. J. (2007). Phage fusion proteins as bioselective receptors for piezoelectric sensors. *ECS Transactions, 2(19),* 9-25.
Onda, T., LaFace, D., Baier, G., Brunner, T., Honma, N., Mikayama, T., Altman, A., & Green, D. R. (1995). A phage display system for detection of T cell receptor-antigen interactions. *Mol. Immunol., 32,* 1387-97.
Orlandi, R., Güssow, D. H., Jones, P. T., & Winter, G. (1989). Cloning immunoglobulin variable domains for expression by the polymerase chain reaction. *Proc. Natl. Acad. Sci. USA, 86,* 3833–7.
Pasqualini, R,. & Ruoslahti, E. (1996). Organ targeting in vivo using phage display peptide libraries. *Nature, 380(6572),* 364-6.
Persson, M.A., Caothien, R. H., & Burton, D. R. (1991). Generation of diverse high-affinity human monoclonal antibodies by repertoire cloning. *Proc. Natl. Acad. Sci. USA, 88,* 2432–6.
Petit, M. A., Jolivet-Reynaud, C., Peronnet, E., Michal, Y., & Trepo, C. (2003). Mapping of a conformational epitope shared between E1 and E2 on the serum-derived human hepatitis C virus envelope. *J. Biol. Chem., 278,* 44385–92.
Petrenko, V. A. (2008a). Evolution of phage display: from bioactive peptides to bioselective nanomaterials. Review. *Expert Opin. Drug Deliv., 5(8),* 1-12.
Petrenko, V. A. (2008b). Landscape phage as a molecular recognition interface for detection devices. *Microelectron. J., 39(2),* 202-7.
Petrenko, V. A., & Smith, G. P. (2000). Phages from landscape libraries as substitute antibodies. *Protein. Eng., 13,* 589-92.

Petrenko, V. A., & Sorokulova, I. B. (2004). Detection of biological threats. A challenge for directed molecular evolution. *J. Microbiol. Methods, 58,* 147-68.

Petrenko, V. A., & Vodyanoy, V. J. (2003). Phage display for detection of biological threat agents. *J. Microbiol. Methods, 53,* 253-62.

Petrenko, V. A., Smith, G. P., Gong, X., & Quinn, T. (1996). A library of organic landscapes on filamentous phage. *Protein. Eng., 9,* 797-801.

Phalipon, A., Folgori, A., Arondel, J., Sgaramella, G., Fortugno, P., Cortese, R., Sansonetti, P. J., & Felici, F. (1997). Induction of anti-carbohydrate antibodies by phage library-selected peptide mimics. *Eur. J. Immunol., 27,* 2620–5.

Ploug, M., Østergaard, S., Gårdsvoll, H., Kovalski, K., Holst-Hansen, C., Holm, A., Ossowski, L., & Danø, K. (2001). Peptide-derived antagonists of the urokinase receptor. Affinity maturation by combinatorial chemistry, identification of functional epitopes, and inhibitory effect on cancer cell intravasation. *Biochemistry, 40(40),* 12157-68.

Poul, M. A., & Marks, J. D. (1999). Targeted gene delivery to mammalian cells by filamentous bacteriophage. *J. Mol. Biol., 288 (2),* 203 -11.

Proba, K., Wörn, A., Honegger, A., & Plückthun, A. (1998). Antibody scFv fragments without disulfide bonds made by molecular evolution. *J. Mol. Biol., 275,* 245–53.

Rader, C., & Barbas, C. F. (1997). Phage display of combinatorial antibody libraries. *Curr. Opin. Biotechnol., 8,* 503-8.

Rajotte, D., Arap, W., Hagedorn, M., Koivunen, E., Pasqualini, R., & Ruoslahti, E. (1998). Molecular heterogeneity of the vascular endothelium revealed by in vivo phage display. *J. Clin. Invest., 102,* 430–7.

Ren, Z., & Black, L. W. (1998). Phage T4 SOC and HOC display of biologically active, full-length proteins on the viral capsid. *Gene, 215,* 439-44.

Roberts, R.W. (1999). Totally in vitro protein selection using mRNA-protein fusions and ribosome display. *Curr. Opin. Chem. Biol., 3(3),* 268-73.

Roberts, R. W., & Szostak, J. W. (1997). RNA-peptide fusions for the in vitro selection of peptides and proteins. *Proc. Natl. Acad. Sci. USA, 94(23),* 12297-302.

Romanov, V. I., Durand, D. B., & Petrenko, V. A. (2001). Phage display selection of peptides that affect prostate carcinoma cells attachment and invasion. *Prostate, 47,* 239-51.

Rosenberg, A., Griffin, K., Studier, F. W., McCormick, M., Berg, J., Novy, R., & Mierendorf, R. (1996). T7Select® phage display system: a powerful new protein display system based on bacteriophage T7. *Innovations (Newsletter of Novagen, Inc.), 6,* 1-6.

Rowley, M. J., Scealy, M., Whisstock, J. C., Jois, J. A., Wijeyewickrema, L. C., & Mackay, I. R. (2000). Prediction of the immunodominant epitope of the pyruvate dehydrogenase complex E2 in primary biliary cirrhosis using phage display. *J. Immunol., 164,* 3413–9.

Saggio, I. & Laufer, R. (1993). Biotin binders selected from a random peptide library expressed on phage. *Biochem. J., 293(3),* 613-6.

Samoylova, T. I., Petrenko, V. A., Morrison, N. E., Globa, L. P., Baker, H. J., & Cox, N. R. (2003). Phage probes for malignant glial cells. *Mol. Cancer Ther., 2,* 1129-37.

Sanghvi, A. B., Miller, K. P.-H., Belcher, A. M., & Schmidt, C. E. (2005). Biomaterials functionalization using a novel peptide that selectively binds to a conducting polymer. *Nature Materials, 4,* 496-502.

Sano, K., & Shiba, K. (2003). A hexapeptide motif that electrostatically binds to the surface of titanium. *J. Am. Chem. Soc., 125,* 14234–5.

Santini, C., Brennan, D., Mennuni, C., Hoess, R. H., Nicosia, A., Cortese, R., & Luzzago, A. (1998). Efficient display of an HCV cDNA expression library as C-terminal fusion to the capsid protein D of bacteriophage lambda. *J. Mol. Biol., 282,* 125-35.

Sarikaya, M., Tamerler, C., Jen, A. K.-Y., Schulten, K., & Baneyx, F. (2003). Molecular biomimetics: nanotechnology through biology. *Nat. Mater., 2,* 577-85.

Sarikaya, M., Tamerler, C., Schwartz, D. T., & Baneyx, F. (2004). Materials assembly and formation using engineered polypeptides. *Annu. Rev. Mater. Res., 34,* 373–408.

Sblattero, D., & Bradbury, A. (2000). Exploiting recombination in single bacteria to make large phage antibody libraries. *Nat. Biotechnol., 18,* 75-80.

Schaffitzel, C., Hanes, J., Jermutus, L., & Plückthun, A. (1999). Ribosome display: an in vitro method for selection and evolution of antibodies from libraries. *J. Immunol. Methods, 231,* 119-35.

Sche, P. P., McKenzie, K. M., White, J. D., & Austin, D. J. (1999). Display cloning: functional identification of natural receptors using cDNA-phage display. *Chem. Biol., 6,* 707-16.

Schimmele, B., & Plückthun, A. (2005). Identification of a functional epitope of the Nogo receptor by a combinatorial approach using ribosome display. *J. Mol. Biol., 352,* 229-41.

Schimmele, B., Gräfe N., & Plückthun, A. (2005). Ribosome display of mammalian receptor domains. *Protein Eng. Des. Sel., 18,* 285-94.

Schlick, T. L., Ding, Z., Kovacs, E. W., & Francis, M. B. (2005). Dual-surface modification of the tobacco mosaic virus. *J. Am. Chem. Soc., 127,* 3718–23.

Schluesener, H. J., & Xianglin, T. (2004). Selection of recombinant phages binding to pathological endothelial and tumor cells of rat glioblastoma by in vivo display. *J. Neurol. Sci., 224(1-2),* 77-82.

Schmidt, C. E., Shastri, V. R., Vacanti, J. P., & Langer, R. (1997). Stimulation of neurite outgrowth using an electrically conducting polymer. *Proc. Natl Acad. Sci. USA, 94,* 8948–53.

Scott, J. K., & Smith, G. P. (1990). Searching for peptide ligands with an epitope library. *Science, 249,* 386-90.

Segal, D. J., Dreier, B., Beerli, R. R., & Barbas, C. F. III (1999). Toward controlling gene expression at will: selection and design of zinc finger domains recognizing each of the 5'-GNN-3' DNA target sequences. *Proc. Natl. Acad. USA, 96,* 2758-63.

Segers, J., Laumonier, C., Burtea, C., Laurent, S., ELst, L. V., & Muller, R. N. (2007). From phage display to magnetophage, a new tool for magnetic resonance molecular imaging. *Bioconjug. Chem. 18(4),* 1251-8.

Sergeeva, A., Kolonin, M. G., Molldrem, J. J., Pasqualini, R., & Arap, W. (2006). Display technologies: application for the discovery of drug and gene delivery agents. *Adv. Drug Deliv. Rev., 58(15),* 1622-54.

Shone, C., Wilton-Smith, P., Appleton, N., Hambleton, P., Modi, N., Gatley, S., & Melling, J. (1985). Monoclonal antibody-based immunoassay for type A *Clostridium botulinum* toxin is comparable to the mouse bioassay. *Appl. Environ. Microbiol., 50,* 63-7.

Siegel, D. L. (2002). Recombinant monoclonal antibody technology. *Transfus. Clin. Biol., 9,* 15-22.

Skelton, N. J., Russell, S., de Sauvage, F., & Cochran, A. G. (2002). Amino acid determinants of beta-hairpin conformation in erythropoeitin receptor agonist peptides derived from a phage display library. *J. Mol. Biol., 316,* 1111–25.

Skerra, A., & Plückthun, A. (1988). Assembly of a functional immunoglobulin Fv fragment in *Escherichia coli. Science, 240,* 1038-41.

Smith, G. P. (1985). Filamentous fusion phage: novel expression vectors that display cloned antigens on the virion surface. *Science, 228(4705),* 1315-7.

Smith, G. P. (1992). Cloning in fUSE vectors. Available from: Prof. G. P. Smith, Division of Biological Sciences, University of Missouri, URL: http://www.biosci.missouri.edu/smithgp/PhageDisplayWebsite/PhageDisplay WebsiteIndex.html.

Smith, G. P., & Petrenko, V. A. (1997). Phage Display. *Chem. Rev., 97,* 391-410.

Sorokulova, I. B., Olsen, E. V., Chen, I. H., Fiebor, B., Barbaree, J. M., Vodyanoy, V. J., Chin, B. A., & Petrenko, V. A. (2005). Landscape phage probes for *Salmonella typhimurium*. *J. Microbiol. Methods, 63,* 55-72.

Souza, G. R., Christianson, D. R., Staquicini, F. I., Ozawa, M. G., Snyder, E. Y., Sidman, R. L., Miller, J. H., Arap, W., & Pasqualini, R. (2006). Networks of gold nanoparticles and bacteriophage as biological sensors and cell-targeting agents. *PNAS, 103,* 1215-20.

Spear, M. A., Breakefield, X. O., Beltzer, J., Schuback, D., Weissleder, R., Pardo, F. S., & Ladner, R. (2001). Isolation, characterization, and recovery of small peptide phage display epitopes selected against viable malignant glioma cells. *Cancer Gene Ther., 8,* 506–11.

Steichen, C., Chen, P., Kearney, J. F., & Turnbough, C. L. Jr. (2003). Identification of the immunodominant protein and other proteins of the *Bacillus anthracis* exosporium. *J. Bacteriol., 185,* 1903–10.

Stemmer, W. P. (1994). Rapid evolution of a protein in vitro by DNA shuffling. *Nature, 370(6488),* 389-91.

Stephen, R. M., & Gillies, R. J. (2007). Promise and progress for functional and molecular imaging of response to targeted therapies. *Pharm. Res. 24(6),* 1172-85.

Stratmann, J., Strommenger, B., Stevenson, K., & Gerlach, G. F. (2002). Development of a peptide-mediated capture PCR for detection of *Mycobacterium avium* subsp. *paratuberculosis* in milk. *J. Clin. Microbiol., 40,* 4244-50.

Tramontano, A., Janda, K. D., & Lerne, R. A. (1986). Catalytic antibodies. *Science, 234,* 1566-70.

Trepel, M., Arap, W., & Pasqualini, R. (2002). In vivo phage display and vascular heterogeneity: implications for targeted medicine. *Curr. Opin. Chem. Biol., 6,* 399–404.

Tseng, R. J., Tsai, C., Ma, L., Ouyang, J., Ozkan, C. S., & Yang, Y. (2006). Digital memory device based on tobacco mosaic virus conjugated with nanoparticles. *Nature Nanotechnology, 1,* 72-7.

Turnbough, C. L. Jr. (2003). Discovery of phage display peptide ligands for species-specific detection of *Bacillus* spores. *J. Microbiol. Methods, 53,* 263-71.

Urry, D. W., McPherson, D. T., Xu, J., Gowda, D. C., & Parker, T. M. (1995). In C. Gebelein, & C. E. Carraher (Eds.), *Industrial Biotechnological Polymers* (pp. 259-281). Lancaster, PA: Technomic.

Valadon, P., Garnett, J. D., Testa, J. E., Bauerle, M., Oh, P., & Schnitzer, J. E. (2006). Screening phage display libraries for organ-specific vascular immunotargeting in vivo. *Proc. Natl. Acad. Sci. USA, 103(2)*, 407-12.

Valentini, R. F., Vargo, T. G., Gardella, J. A. Jr., & Aebischer, P. (1992). Electrically conductive polymeric substrates enhance nerve fibre outgrowth in vitro. *Biomater., 13*, 183–90.

vanZonneveld, A. J., vandenBerg, B. M. M., vanMeijer, M., & Pannekoek, H. (1995). Identification of functional interaction sites on proteins using bacteriophage-displayed random epitope libraries. *Gene, 167*, 49-52.

Vidal, J. C., Garcia, E., & Castillo, J. R. (1999). In situ preparation of a cholesterol biosensor: entrapment of cholesterol oxidase in an overoxidized polypyrrole film electrodeposited in a flow system: Determination of total cholesterol in serum. *Anal. Chim. Acta., 385*, 213–22.

Villemagne, D., Jackson, R., & Douthwaite, J. A. (2006). Highly efficient ribosome display selection by use of purified components for in vitro translation. *J. Immunol. Methods, 313*, 140-8.

Wan, J., Johnson, M. L., Guntupalli, R., Petrenko, V. A., & Chin, B. A. (2007a). Detection of *Bacillus anthracis* spores in liquid using phage-based magnetoelastic micro-resonators. *Sensors Actuators B, 127*, 559-66.

Wan, J., Shu, H., Huang, S., Fiebor, B., Chen, I. H., Petrenko, V. A., & Chin, B. A. (2007b). Phage-based magnetoelastic wireless biosensors for detecting *Bacillus anthracis* spores. *IEEE Sensors J., 7(3)*, 470-7.

Wang, C. I., Yang, Q., & Craik, C. S. (1996). Phage display of proteases and macromolecular inhibitors. *Combinatorial Chemistry, 267*, 52-68.

Wang, L. F., & Yu, M. (2004). Epitope identification and discovery using phage display libraries: applications in vaccine development and diagnostics. *Curr. Drug Targets, 5*, 1–15.

Wang, L. F., Duplessis, D. H., White, J. R., Hyatt, A.D., & Eaton, B. T. (1995). Use of a gene-targeted phage display random epitope library to map an antigenic determinant on the Bluetongue Virus outer capsid protein Vp5. *J. Immunol. Methods, 178*, 1-12.

Weaver-Feldhaus, J. M., Lou, J., Coleman, J. R., Siegel, R. W., Marks, J. D., & Feldhaus, M.J. (2004). Yeast mating for combinatorial Fab library generation and surface display. *FEBS Lett., 564(1-2)*, 24-34.

Weissleder, R. (2006). Molecular imaging in cancer. *Science, 312(5777)*, 1168-71.

Whaley, S. R., English, D. S., Hu, E. L., Barbara, P. F., & Belcher, A. M. (2000). Selection of peptides with semiconductor binding specificity for directed nano-crystal assembly. *Nature, 405*, 665-8.

Willats, W. G. (2002). Phage display: practicalities and prospects. *Plant Mol. Biol., 50,* 837–54.

Wilson, D. R., & Finlay, B. B. (1998). Phage display: application, innovations, and issues in phage and host biology. *Can. J. Microbiol., 44,* 313–29.

Wilson, D. S., Keefe, A. D., & Szostak, J. W. (2001). The use of mRNA display to select high-affinity protein-binding peptides. *Proc. Natl. Acad. Sci. USA, 98,* 3750-5.

Wolfe, S. A., Greisman, H. A., Ramm, E. I., & Pabo, C.O. (1999). Analysis of zinc fingers optimized via phage display: evaluating the utility of a recognition code. *J. Mol. Biol., 285,* 1917-34.

Worn, A., & Plückthun, A. (2001). Stability engineering of antibody single-chain Fv fragments. *J. Mol. Biol., 305,* 989–1010.

Wrighton, N. C., Farrell, F. X., Chang, R., Kashyap, A. K., Barbone, F. P., Mulcahy, L. S., Johnson, D. L., Barrett, R. W., Jolliffe, L. K., & Dower, W. J. (1996). Small peptides as potent mimetics of the protein hormone erythropoietin. *Science, 273,* 458–64.

Wu, H., Yang, W. P., & Barbas, C. F. (1995). Building zinc fingers by selection - toward a therapeutic application. *Proceedings of the National Academy of Sciences of the United States of America, 92,* 344-8.

Xu, L., Aha, P., Gu, K., Kuimelis, R. G., Kurz, M., Lam, T., Lim, A. C., Liu, H., Lohse, P. A., Sun, L., Weng, S., Wagner, R. W., & Lipovsek, D. (2002). Directed evolution of high-affinity antibody mimics using mRNA display. *Chem. Biol., 9,* 933-42.

Yang, L. M., Diaz, J. E., McIntire, T. M., Weiss, G. A., & Penner, R. M. (2008). Covalent virus layer for mass-based biosensing. *Anal. Chem., 80(4),* 933-43.

Yang, L. M., Tam, P. Y., Murray, B. J., McIntire, T. M., Overstreet, C. M., Weiss, G. A., & Penner, R. M. (2006). Virus electrodes for universal biodetection. *Anal. Chem., 78(10),* 3265-70.

Yoo, P. J., Nam, K. T., Qi, J., Lee, S.-K., Park, J., Belcher, A. M., & Hammond, P. T. (2006). Spontaneous assembly of viruses on multilayered polymer surfaces. *Nature Materials, 5(3),* 234-40.

Yu, J., & Smith, G. P. (1996). Affinity maturation of phage-displayed peptide ligands. *Methods Enzymol., 267,* 3-27.

Yuan, L., Kurek, I., English, J., & Keenan, R. (2005). Laboratory-directed protein evolution. *Microbiol. Mol. Biol. Rev., 69(3),* 373-92.

Zhang, L., Hoffman, J. A., & Ruoslahti, E. (2005). Molecular profiling of heart endothelial cells. *Circulation, 112(11),* 1601-11.

INDEX

A

accounting, 48
accuracy, 3, 35
acetonitrile, 39
acetylcholine, 66, 74
acid, 9, 27, 42, 47, 59, 71, 86
acidic, 45
acoustic, 36, 82
ACS, 20
active site, 8, 11
acute, 37
acute myeloid leukemia, 37
addiction, 57
adenosine, 17
adhesion, 69
administration, 63
adsorption, 31, 39, 40, 82
agar, 32
agent, 63
agents, 3, 32, 35, 36, 39, 42, 45, 62, 63, 76, 77, 83, 85, 86
agonist, 79, 80, 86
alcohol, 39
alkaline, 39
alternative, 1, 20, 38, 52, 58
alternatives, 55, 78
amide, 17
amine, 59
amines, 59
amino, 9, 18, 24, 27, 38, 40, 42, 48
amino acid, 9, 18, 24, 27, 42, 48
amino acids, 18, 24, 48
amino-groups, 38
angiogenesis, 62
angiotensin-converting enzyme, 76
animals, 32
antagonists, ii, 27, 76, 81, 83
antiangiogenic, 72
anti-bacterial, 58
antibiotics, 55, 58
anticancer, 58
anticoagulants, 72
antigen, 30, 31, 32, 57, 58, 68, 71, 73, 79, 81, 82
antigen presenting cells, 57
anti-tumor, 56, 58
apoptosis, 56, 58, 62
apoptotic cells, 63, 73
application, 3, 9, 16, 18, 41, 44, 50, 52, 72, 85, 88
arginine, 6
aromatic rings, 50
aspartate, 44
assessment, 67
asymptomatic, 69
attachment, 48, 51, 62, 84
availability, 38

B

B cell, 30
Bacillus, 68, 77, 86, 87
Bacillus subtilis, 77
bacteria, 3, 20, 40, 46, 50, 55, 61, 85
bacterial, 15, 17, 18, 20, 24, 26, 36, 41, 45, 80
bacterial cells, 15, 24, 41
bacteriophages, 23, 55, 78
bacterium, 52
barrier, 50
bioassay, 86
biocatalysts, 70
biocompatible, 46
biological activity, 6
biological processes, 62
biological systems, 1, 8
biomaterial, 4, 51
biomaterials, 3, 50, 55
biomedical applications, 16, 47, 51
biomolecule, 51
biomolecules, 4, 47, 48
biophysics, 52
biopolymer, 52
biosensors, 36, 39, 51, 65, 73, 82, 87
biotechnological, 2
biotechnology, x, ii, 2, 4, 8, 55, 71
blood, 58
blood vessels, 58
bonding, 68, 82
bonds, 26, 83
bottom-up, 45
botulinum, 73
brain, 57
buffer, 45
building blocks, 45

C

cancer, 16, 36, 58, 62, 63, 76, 78, 83, 88
cancer cells, 58, 62
cancer treatment, 62
carbohydrate, 56, 72, 83
carbon, 8, 73
carbon nanotubes, 8, 73
carboxyl, 40
carboxylic, 59
carcinoma, 82, 84
cargo, 58
carrier, 38, 44, 81
cassettes, 30
catalysis, 16, 32
catalytic activity, 65
cell differentiation, 56
cell line, 26, 37
cell surface, 18, 19, 78
central nervous system, 57, 72
charge trapping, 50
chelating agents, 15
chemical approach, 5
chemicals, 8
chimera, 11, 24
chloride, 51
chlorine, 50
cholesterol, 87
chromatography, 14
circulation, 62
clinical trial, 29
clinical trials, 29
clone, 44, 59
cloning, 24, 25, 70, 83, 85
coatings, 51
cobalt, 49
cocaine, 57
coding, 17, 18, 24, 25, 26
codon, 14
colon cancer, 77
complementarity, 30
complementary DNA, 1
composites, 39
composition, 44, 48
compounds, 6, 38
concentration, 37, 45
condensation, 6
conductance, 50
conductive, 46, 50, 87
conductivity, 50, 51
conjugation, 59, 60
consensus, 66, 68, 77

constraints, 46
contingency, 6, 7
control, 7, 47, 57, 63
coupling, 29, 47, 51, 59, 63
covalent, 14, 17, 18, 40, 43, 59, 60
covering, 28
cross-linking, 31
crystal growth, 49, 69
crystalline, 48, 49
crystals, 47
culture, 59
cycles, 16
cytometry, 3, 20

D

definition, 27
degenerate, 25
degradation, 18, 57, 60
degrading, 15
dehydrogenase, 84
dendritic cell, 71
density, 10
deposition, 71
derivatives, 59
detachment, 52, 72
detection, 2, 3, 33, 35, 36, 37, 38, 41, 47, 55, 66, 72, 76, 78, 82, 83, 86, 87
diabetes, 81
differentiation, 56
diffusion, 46
digestion, 7
discipline, 9, 43
discrimination, 63
diseases, 36
dissociation, 15
disulfide, 25, 31, 40, 59, 80, 81, 83
disulfide bonds, 83
diversification, 13, 15, 16
diversity, 13, 15, 18, 30, 32, 55, 58, 66, 74
donor, 30
donors, 37, 50
dopant, 51
double helix, 11
drug delivery, 51, 78

drugs, 3, 19, 55
durability, 38

E

economics, 9
electrical conductivity, 50
electrodes, 89
encapsulated, 60
encapsulation, 61
encoding, 10, 79, 81
endothelial cell, 89
endothelium, 84
end-users, 44
energy, 2, 8, 10, 50
engines, 68
entrapment, 87
environment, 3, 8, 35
environmental conditions, 6, 38
enzymatic, 19
enzymes, 5, 7, 11, 14, 27, 35, 76
epitopes, 25, 27, 32, 36, 45, 74, 81, 83, 86
equilibrium, 49
erythropoietin, 80, 88
exploitation, 79
exposure, 39

F

fabricate, 8, 42, 44
family, 11, 67, 81
family members, 11
fibers, 44, 67
fibroblast growth factor, 60
fibronectin, 14, 78
filament, 52
film, 51, 87
film thickness, 51
fitness, 26
floating, 46
flow, 3, 20, 87
fluorescence, 36, 38, 46, 73
folding, 6, 7, 71, 72, 74, 78
food, 3, 8, 35

freedom, 37
fullerene, 8, 70, 73
functionalization, 50, 52, 84
fusion, 14, 19, 23, 25, 37, 38, 44, 45, 56, 59, 60, 69, 82, 84, 86
fusion proteins, 82

G

gene expression, 85
gene pool, 29
gene therapy, 55, 60, 67, 71
gene transfer, 32, 77
genes, 11, 32, 61, 81
genetic information, 15, 23, 29
genome, 18, 24, 43, 56, 79
genomics, 2, 28
genotype, 15, 23
glial cells, 84
glioblastoma, 85
glioma, 61, 86
glutamate, 44
glycine, 6
glycosylation, 20
goals, 8, 44, 45
gold, 36, 40, 45, 46, 49, 59, 86
gold nanoparticles, 86
groups, 38, 47, 51
growth, 44, 47, 49, 56, 60, 69, 72
guanine, 50
guidance, 51

H

harvesting, 50
health, 8
hearing, 72
heart, 89
helix, 7, 11, 69
hepatitis B, 80, 81
hepatitis C, 83
heterogeneity, 84, 87
high temperature, 39
hips, 37

hormone, 79, 88
hormones, 27
host, 26, 30, 46, 49, 52, 57, 88
human, 8, 13, 29, 30, 61, 66, 69, 70, 71, 73, 77, 80, 81, 82, 83
human immunodeficiency virus, 69
humans, 30
hybrid, 17, 50, 52, 61, 75
hybridization, 15, 61
hybridoma, 30, 32
hybrids, 14
hydrolyzed, 7
hydrophobic interactions, 52

I

identification, 3, 26, 68, 77, 83, 85, 88
immobilization, 38, 39, 40, 51
immune response, 30, 58, 81
immunity, 32
immunization, 71, 80
immunoassays, 3
immunogen, 30, 45
immunogenicity, 58
immunoglobulin, 16, 70, 83, 86
immunoglobulin superfamily, 16
inactivation, 32
incubation, 31
industrial, 42
industry, 4, 55
infection, 52, 80
infrared, 46
inhibition, 55, 80
inhibitor, 80
inhibitory effect, 83
injection, 60
inorganic, 4, 40, 46, 47, 48, 49, 53
insertion, 6
integration, 47
integrin, 60, 61
interaction, 18, 31, 42, 44, 58, 87
interactions, 2, 19, 23, 47, 48, 50, 52, 56, 82
interface, 46, 62, 83
interleukin, 69
internalization, 46, 60

invasive, 77
iodine, 51
ionic, 50
ions, 49, 51
iron, 63, 69
isolation, 20
iteration, 13

K

kinase, 19, 71
kinases, 62
King, 74

L

label-free, 41
labeling, 41
lambda, 57, 72, 77, 80, 84
landscapes, 83
leukemia, 37
leukocytes, 37
life forms, 1
ligand, 14, 20, 27, 42, 45, 56, 62, 75, 79, 82
linear, 19, 25
linkage, 44
links, 2
lipase, 80
lipid, 56, 59
liposomes, 59, 82
liver, 57, 63, 79
lyme disease, 78
lymphocytes, 36

M

magnetic, x, 14, 19, 25, 31, 49, 50, 63, 80, 85
magnetic beads, 15, 31
magnetic resonance, 63, 85
magnetic resonance imaging, 63
magnetoelastic, 36, 39, 77, 78, 87
maintenance, 61
malignant, 84, 86
Mammalian, 72

mammalian cell, 20, 58, 60, 61, 76, 77, 78, 83
mammals, 60
manipulation, 38, 46, 55
manufacturing, 55
mapping, 27, 66, 71, 79
markets, 42
mathematics, 9
matrix, 31
maturation, 31, 32, 76, 83, 89
medical diagnostics, 43
medicine, 78, 87
melanoma, 69
membranes, 59
memory, 50, 87
messages, 53
metagenomics, 65
metal ions, 49
metal oxide, 47
metals, 49
metastasis, 62
mice, 57, 81, 82
microarray, 15, 37, 65, 67
microbial, 40, 65
microorganism, 18, 41
microorganisms, 20, 58, 74
microscopy, 46
milk, 87
mimicking, 16, 56
mimicry, 79
model system, 63
modeling, 1
moieties, 52
molecular markers, 62
molecular structure, 5
monoclonal, 32, 36, 39, 40, 69, 70, 71, 73, 83, 86
monoclonal antibodies, 32, 36, 40, 69, 70, 71, 83
monoclonal antibody, 32, 39, 73, 86
monolayer, 46
morphology, 49
mosaic, 49, 74, 85, 87
mouse, 36, 86
mRNA, 13, 14, 15, 16, 17, 18, 19, 66, 67, 71, 74, 75, 79, 82, 84, 88, 89

mutagenesis, 9, 10, 13, 15, 18, 33
mutagenic, 20, 31
mutants, 10, 69
mutation, 10
mutations, 9, 10, 11, 15, 16, 32, 71
myasthenia gravis, 74
myeloid, 37

N

nanobiotechnology, 8, 41
nanocrystals, 48, 49
nano-crystals, 3
nanomaterials, 44, 52, 83
nanoparticles, 3, 46, 47, 50, 82, 86, 87
nanostructured materials, 43
nanostructures, 3, 45, 47
nanosystems, 45
nanotechnology, 43, 47, 53, 66, 85
nanotubes, 4, 8, 44, 49, 73
nanowires, 4, 46, 49, 80
natural evolution, 10, 23, 47
nerve, 51, 87
nerve regeneration, 51
next generation, 26
nickel, 40
non-invasive, 62
normal, 36, 37, 61, 63
nucleation, 52
nucleic acid, i, 5, 35, 71
nucleotide sequence, 14

O

oat, 25, 59, 74, 80
oligonucleotides, 25
one dimension, 44
optical, 46, 47, 52, 82
optical imaging, 82
optical properties, 46, 47
optical tweezers, 52
optimization, 61
organ, 62, 87
organic, 4, 35, 39, 46, 47, 83

organic solvent, 39, 47
organism, 10, 18
orientation, 37, 40, 48, 52
oxide, 63, 69

P

packaging, 25, 26
pancreatic, 80
particles, 23, 24, 29, 40, 45, 52, 57, 60, 61, 63, 75, 77, 79, 80
pathogenic, 2, 55, 74
pathogenic agents, 2
pathogens, 35
patients, 37, 62, 63, 69, 74, 78
PCR, 11, 13, 14, 33, 86
peptide bonds, 26
perchlorate, 51
periodic, 7
periodicity, 49
permit, 7
phagocytic, 58, 63
pharmaceutical, 59
phenotype, 15, 23
philosophical, 1
phosphatidylserine, 73
photosynthetic, 50
physical properties, 6, 7
physicochemical, 11
physico-chemical characteristics, 44
physicochemical properties, 11
physics, 1
piezoelectric, 36, 82
pitch, 24
plasmid, 57
plastic, 31, 39, 40, 53, 65
platforms, 33
platinum, 50
play, 3
polyelectrolytes, 46
polymer, 45, 46, 48, 50, 67, 71, 84, 85, 89
polymerase, 3, 30, 83
polymerase chain reaction, 3, 30, 83
polymeric materials, 80
polymers, 35, 48, 51

Index

polypeptides, 7, 15, 18, 19, 20, 24, 49, 69, 85
polystyrene, 25, 48
pools, 15
poor, 18, 47
population, 7, 26, 44
preclinical, 75
prediction, 61
pressure, 15, 18
primary biliary cirrhosis, 84
prior knowledge, 56
probe, 1, 3, 35, 38, 39, 42, 63, 66, 82
production, 38, 46, 55, 59
programming, 43
prokaryotic, 17
proliferation, 56
property, x, 7, 39, 49, 59
prophylaxis, 67
prostate, 79, 82, 84
prostate carcinoma, 82, 84
proteases, 6, 31, 62, 88
protein arrays, 13
protein binding, 25
protein design, 7, 8, 9, 73
protein engineering, i, 5, 9, 23, 28
protein folding, 78
protein function, 11, 24
protein kinases, 62
protein sequence, 7, 57, 65
protein structure, i, 2, 5, 16
proteolytic enzyme, 6
proteomes, 75
proteomics, 2, 28, 41, 42, 76, 79
protocol, 17
pseudo, 49
pseudomonas aeruginosa, 69
purification, 3, 40, 59
pyruvate, 84

Q

quality control, 20
quantum, 78, 80
quantum dots, 78
quartz, 36, 72

R

radio, 63
range, 16, 27, 36, 48, 57
rat, 61, 85
reaction time, 41
reactivity, 33
reagent, 41, 63
reagents, 2, 3, 19, 35, 58, 74
receptor agonist, 27, 80, 86
recognition, 35, 37, 39, 40, 41, 42, 48, 49, 52, 62, 69, 76, 77, 82, 83, 88
recombinant DNA, 6, 11, 24, 26
recombination, 11, 18, 31, 85
recovery, 75, 86
reengineering, 59
refractory, 16
regeneration, 51
regenerative medicine, 56
relevance, 52
repair, 43
replicability, 26
reporters, 46
reproduction, 11
reservoir, 59
residues, 7, 24, 44, 57
resin, 14
rheological properties, 6
ribosomal, 14
ribosome, 13, 14, 15, 16, 17, 18, 65, 75, 79, 82, 84, 85, 87
ribosomes, 15, 17
rings, 50
RNA, 17, 27, 50, 67, 72, 84
robotic, 11
robustness, 39
room temperature, 39

S

salt, 15, 45
sample, 41
scaffold, 46, 62, 78
scaffolding, 8

scaffolds, 14, 44
scattering, 46
search, 80
searches, ii, 47
secretion, 67
selecting, 19, 32, 44, 49, 50, 73, 79
selectivity, 4, 13, 36, 47
self-assembly, 39, 43
self-organizing, 49
semiconductor, 4, 47, 48, 49, 52, 73, 88
semiconductors, 48
sensing, 38, 79
sensitivity, 38
sensors, 3, 10, 36, 39, 82, 86
separation, 3, 15, 19, 37
series, 44
serum, 56, 74, 83, 87
services, x
sexual reproduction, 11
shape, 47
shipping, 39
side effects, 58, 63
signal peptide, 56
signaling, 19
signalling, 32
signals, 37, 47
silk, 44
simulation, 1
sites, 3, 11, 24, 75, 87
solid surfaces, 40
solid tumors, 62
solubility, 6, 47
solvent, 24
sorting, 20, 73
species, 40, 52, 77, 87
specificity, 7, 10, 33, 36, 38, 48, 51, 58, 73, 88
spectroscopy, 46
speed, 3
spleen, 30
stability, 3, 6, 7, 16, 17, 19, 38, 51, 58, 77
stages, 14
standardization, 38
stem cell differentiation, 56
steric, 40, 46
storage, 39, 45, 50

strain, 68
strains, 33
strategies, 9, 13, 33, 61, 75, 80
strength, 50, 51
streptavidin, 31, 45, 59, 63
stretching, 37, 77
structural knowledge, 9, 10
structural protein, 23
styrene, 51
substances, 56
substitutes, 37
supramolecular, 43
surface chemistry, 43
surface modification, 76, 85
surface properties, 44
surface structure, 26
surrogates, 3, 38
survival, 26
susceptibility, 38
switching, 49, 50
symmetry, 44
synthesis, xi, 2, 3, 5, 6, 38, 47, 49, 51, 80
synthetic polymers, 48

T

T cell, 82
target identification, 3
technology, 1, 2, 4, 11, 16, 17, 18, 20, 23, 24, 26, 35, 40, 44, 45, 59, 72, 75, 86
temperature, 14
ternary complex, 14
therapeutic gene delivery, 60
therapeutic targets, 2
therapeutics, 55, 76
thermodynamic stability, 7
thermostability, 10, 39
threats, 33, 83
three-dimensional, 41
thrombin, 6
thymus, 74
time, 2, 3, 15, 36, 41
tissue, 26, 43, 57, 60, 61, 75
titanium, 50, 84
tobacco, 49, 85, 87

total cholesterol, 87
toxic, 32, 38, 60
toxin, 86
transcriptional, 56
transducer, 73
transduction, 61
transfection, 72
transfer, 32, 77, 78
transformation, 15, 18, 61
transgene, 61, 63
translation, 16, 17, 87
transport, 48, 58
tripeptide, 6
tryptophan, 44
tubular, 23, 49
tumor, 26, 37, 58, 62, 68, 69, 75, 78, 85
tumor cells, 69, 85
tumors, 58, 62, 63, 79
tumour, 47
two-dimensional, 46, 77
tyrosine, 17, 71

U

underlying mechanisms, i
uniform, 45
urea, 39

urokinase, 83

V

validation, 3, 62, 79
variability, 38
vascular endothelial growth factor, 72
vasculature, 58, 66, 68
vehicles, 51, 55, 56, 58, 60
versatility, 8, 60, 61
vessels, 58
virulence, 41
virus, 36, 44, 45, 49, 57, 61, 69, 74, 81, 82, 83, 85, 87, 89
viruses, 3, 43, 44, 46, 49, 50, 60, 71, 76, 78, 89
viscosity, 6
visualization, 62

W

warfare, 3, 42
water, 6, 44
wireless, 36, 77, 87
workers, 45